Cloud Computing Enabled Big-Data Analytics in Wireless Ad-hoc Networks

Wireless Communications and Networking Technologies: Classifications, Advancement and Applications

Series Editor
D.K. Lobiyal, R.S. Rao and Vishal Jain

The series addresses different algorithms, architecture, standards and protocols, tools and methodologies which could be beneficial in implementing next generation mobile network for the communication. Aimed at senior undergraduate students, graduate students, academic researchers and professionals, the proposed series will focus on the fundamentals and advances of wireless communication and networking, and their such as mobile ad-hoc network (MANET), wireless sensor network (WSN), wireless mess network (WMN), vehicular ad-hoc networks (VANET), vehicular cloud network (VCN), vehicular sensor network (VSN) reliable cooperative network (RCN), mobile opportunistic network (MON), delay tolerant networks (DTN), flying ad-hoc network (FANET) and wireless body sensor network (WBSN).

Cloud Computing Enabled Big-Data Analytics in Wireless Ad-hoc Networks
Sanjoy Das, Ram Shringar Rao, Indrani Das, Vishal Jain and Nanhay Singh

Smart Cities
Concepts, Practices, and Applications
Krishna Kumar, Gaurav Saini, Duc Manh Nguyen, Narendra Kumar and Rachna Shah

For more information about this series, please visit:
https://www.routledge.com/Wireless%20Communications%20and%20Networking%20Technologies/book-series/WCANT

Cloud Computing Enabled Big-Data Analytics in Wireless Ad-hoc Networks

Edited by
Sanjoy Das, Ram Shringar Rao, Indrani Das,
Vishal Jain, and Nanhay Singh

CRC Press
Taylor & Francis Group
Boca Raton London New York

CRC Press is an imprint of the
Taylor & Francis Group, an **informa** business

First edition published 2022
by CRC Press
6000 Broken Sound Parkway NW, Suite 300, Boca Raton, FL 33487-2742

and by CRC Press
2 Park Square, Milton Park, Abingdon, Oxon, OX14 4RN

© 2022 selection and editorial matter, Sanjoy Das, Ram Shringar Rao, Indrani Das, Vishal Jain, and Nanhay Singh; individual chapters, the contributors

First edition published by CRC Press 2022

CRC Press is an imprint of Taylor & Francis Group, LLC

Reasonable efforts have been made to publish reliable data and information, but the author and publisher cannot assume responsibility for the validity of all materials or the consequences of their use. The authors and publishers have attempted to trace the copyright holders of all material reproduced in this publication and apologize to copyright holders if permission to publish in this form has not been obtained. If any copyright material has not been acknowledged please write and let us know so we may rectify in any future reprint.

Except as permitted under U.S. Copyright Law, no part of this book may be reprinted, reproduced, transmitted, or utilized in any form by any electronic, mechanical, or other means, now known or hereafter invented, including photocopying, microfilming, and recording, or in any information storage or retrieval system, without written permission from the publishers.

For permission to photocopy or use material electronically from this work, access www.copyright.com or contact the Copyright Clearance Center, Inc. (CCC), 222 Rosewood Drive, Danvers, MA 01923, 978-750-8400. For works that are not available on CCC please contact mpkbookspermissions@tandf.co.uk

Trademark notice: Product or corporate names may be trademarks or registered trademarks and are used only for identification and explanation without intent to infringe.

ISBN: 978-0-367-75442-6 (hbk)
ISBN: 978-1-032-07328-6 (pbk)
ISBN: 978-1-003-20645-3 (ebk)

DOI: 10.1201/9781003206453

Typeset in Times
by codeMantra

Contents

Preface vii
Editors ix
Contributors xi
About This Book xv

Chapter 1 Cloud-Based Underwater Ad-hoc Communication: Advances, Challenges, and Future Scopes 1

    Rinki Chauhan, Ram Shringar Rao, and Sanjoy Das

Chapter 2 A Hybrid Cryptography Technique with Blockchain for Data Integrity and Confidentiality in Cloud Computing 15

    K.L. Neela and R.K. Ramesh

Chapter 3 Fog Computing Environment in Flying Ad-hoc Networks: Concept, Framework, Challenges, and Applications 31

    Devraj, Ram Shringar Rao, and Sanjoy Das

Chapter 4 Wi-Fi Computing Network Empowers Wi-Fi Electrical Power Network 49

    Y.P. Chawla

Chapter 5 Big Data Analytics for Vehicular Edge Networks 65

    Jayashree Patil, Suresha, and Nandini Sidnal

Chapter 6 Impact of Various Parameters on Gauss Markov Mobility Model to Support QoS in MANET 85

    Munsifa Firduas Khan and Indrani Das

Chapter 7 Heterogeneous Ad-hoc Network Management: An Overview 103

    Mehajabeen Fatima and Afreen Khursheed

Chapter 8 Deployment of the Biometrics-as-a-Service (BaaS) Design for the Internet of Biometric Things (IoBT) on the AWS Cloud 125

    Vinayak Ashok Bharadi, Trupti S. Kedar, Pravin S. Jangid, and Mamta Meena

Chapter 9 A Comprehensive Survey of Geographical Routing in Multi-hop Wireless Networks 141

Allam Balaram, Manda Silparaj, Shaik Abdul Nabi, and P. Chandana

Chapter 10 Energy-Aware Secure Routing in Sensor Network 173

N. Ambika

Chapter 11 Deploying Trust-Based E-Healthcare System Using Blockchain-IoT in Wireless Networks... 189

Amanjot Kaur and Parminder Singh

Chapter 12 Low Cost Robust Service Overloading Fusion Model for Cloud Environments 203

Sitendra Tamrakar, Ramesh Vishwakarma, and Sanjeev Kumar Gupta

Chapter 13 Load Balancing Based on Estimated Finish Time of Services........ 217

Rajesh Sachdeva

Chapter 14 Blockchain-Enabled Smart Contract Optimization for Healthcare Monitoring Systems 229

Nitima Malsa, Vaibhav Vyas, and Pooja Singh

Chapter 15 Interference Mitigation Using Cognitive Femtocell from 5G Perspective 251

Gitimayee Sahu and Sanjay S. Pawar

Index..269

Preface

In recent times, tremendous growth and usage of various wireless ad-hoc networks such as Mobile Ad-hoc Networks (MANETs), Vehicular Ad-hoc Networks (VANETs), Wireless Sensor Networks (WSNs), and IoT-based Cloud Networks (ICN) has emerged. These emerging networks open lots of opportunities for researchers to explore these areas and enhance their efficacy. Advanced technologies like cloud computing, Big-data, and Internet of Things (IoT)-based services have been well incorporated in wireless networks. Day by day, wireless devices and their uses are increasing very rapidly. Many industries are developing huge numbers of applications of wireless networks to ease our day-to-day works; simultaneously, these applications generate a large volume of data. To better manage these data and optimize network performance, IoT, cloud, and Big-data techniques play a vital role. Optimizing traffic management operations is a big challenge globally due to a massive increase in the number of vehicles, traffic congestions, and road accidents. Therefore, to minimize a large number of traffic accidents, improve the performance of transportation systems, enhance road safety, improve traffic efficiency, and protect the environment, VANETs play an essential role. VANET has been developed for Intelligent Transportation Systems (ITS), and it is the future of road transportation systems. Current development in wireless communication, computing paradigms, Big-data, and cloud computing fetched the enhancement of intelligent devices equipped with wireless communication capabilities and high-performance processing tools. Various applications of vehicular communication systems generate a large amount of data. Nowadays, these data are placed on a cloud called cloud data and described as Big-data. Researchers and data scientists using IoT include rapidly growing data that can be handled by using Big-data. Big-data refers to many structured and unstructured data that can be analyzed using various application software. The Big-data on the cloud will have profound impacts on the implementation of VANETs, which makes transportation systems much safer, efficient, and effective. The integration of edge computing with VANETs performs data analytics. Edge computing technology has various benefits in ad-hoc and vehicular networks in terms of efficient bandwidth, minimum latency, and context awareness. Since edge nodes have lesser data storage capability, the data will be moved to cloud servers for future uses. These data are available all the time as and when required; this helps in improving service availability. Routing in VANET based on location-aware routing is done with the help of the cloud with more precision.

Cloud computing and Big-data include the various heterogeneous technologies that work together and provide real-time solutions for the Flying Ad-hoc Network (FANET), MANET, VANET, etc. Some cloud service providers offer cloud-based services for vehicular node users on a payment basis, and some are free services. With the development of wireless technologies, Big-data, IoT, and cloud computing and rapid growth in the number of intelligent vehicles, the need to connect intelligent devices such as smartphones, PDAs, smartwatch, smart TV, and laptop with the cloud through the Internet has increased and is high on demand. Further, all types of wireless networks face several technical challenges in deployment due to less flexibility, poor connectivity, less fixed infrastructure, and inadequate intelligence. Therefore, cloud computing, Big-data analytics, and IoT manage various wireless ad-hoc networks very efficiently and effectively.

Wireless networks, especially in a heterogeneous environment and their related issues, emerge as a very challenging and open future scope of research. Nowadays, sensor networks deployed for real-time monitoring of houses, building, paddy field, patient, etc. simultaneously process vast amounts of data to efficiently handle network performance. These help in timely execute appropriate actions based on processed information. Therefore, to overcome these issues, the cloud- and Big-data-based applications that support various networks are essential for achieving high accuracy, availability, and real-time data analysis, which may not be resolved by traditional network infrastructure. Hence, integration of cloud computing, Big-data, and IoT for various types of networks has been included.

Editors

Dr. Sanjoy Das is currently working as a Professor in the Department of Computer Science, Indira Gandhi National Tribal University (a Central Government University), Amarkantak, M.P. (Manipur Campus), India. He did his B.E. and M.Tech. in Computer Science and Engineering. He has obtained his Ph.D. in Computer Science and Technology from the School of Computer and Systems Sciences, Jawaharlal Nehru University, New Delhi, India. Before joining IGNTU, he has worked as an Associate Professor in the School of Computing Science and Engineering, Galgotias University, India from July 2016 to September 2017. He has worked as an Assistant Professor at Galgotias University from September 2012 to June 2016 and as an Assistant Professor at G. B. Pant Engineering College, Uttarakhand, and Assam University, Silchar, from 2001 to 2008. His current research interest includes Mobile Ad-hoc Networks, Vehicular Ad-hoc Networks, Distributed Systems, and Data Mining. He has published numerous papers in international journals and conferences, including IEEE, IGI-Global, Inderscience, Elsevier, and Springer.

Dr. Ram Shringar Rao is currently working as an Associate Professor in the Department of Computer Science and Engineering at Netaji Subhas University of Technology, East Campus, Delhi, India. He has received his B.E. (Computer Science and Engineering) and M.Tech. (Information Technology) in 2000 and 2005, respectively. He has obtained his Ph.D. (Computer Science and Technology) from the School of Computer and Systems Sciences, Jawaharlal Nehru University, New Delhi, India in 2011. He has worked as an Associate Professor in the Department of Computer Science, Indira Gandhi National Tribal University (a Central University, MP) from April 2016 to March 2018. He has more than 18 years of teaching, administrative, and research experience. Currently, he is associated with a wide range of journals and conferences as chief editor, editor, chair, and member. His current research interest includes Mobile Ad-hoc Networks and Vehicular Ad-hoc Networks. He is supervising many M.Tech. and Ph.D. students. Dr. Rao has published more than 100 research papers with good impact factors in reputed international journals and conferences, including IEEE, Elsevier, Springer, Wiley & Sons, Taylor & Francis, Inderscience, Hindawi, IERI Letters, and American Institute of Physics.

Dr. Vishal Jain, Ph.D., is an Associate Professor in the Department of Computer Science and Engineering, School of Engineering and Technology, Sharda University, Greater Noida, UP, India. Before that, he has worked for several years as an Associate Professor at Bharati Vidyapeeth's Institute of Computer Applications and Management (BVICAM), New Delhi. He has more than 14 years of experience in the academics. He has more than 480 research citation indices with Google Scholar (h-index score 12 and i-16 index 11). He has authored more than 85 research papers in reputed conferences and journals, including the Web of Science and Scopus. He has authored and edited more than 15 books with various reputed publishers, including

Elsevier, Springer, Apple Academic Press, CRC, Taylor and Francis Group, Scrivener, Wiley, Emerald, DeGruyter, and IGI-Global. His research areas include information retrieval, semantic web, ontology engineering, data mining, ad-hoc networks, and sensor networks. He received a Young Active Member Award for the year 2012–2013 from the Computer Society of India, a Best Faculty Award for the year 2017, and Best Researcher Award for the year 2019 from BVICAM, New Delhi. He holds Ph.D. (CSE), M.Tech. (CSE), M.B.A. (HR), M.C.A., M.C.P., and C.C.N.A.

Dr. Indrani Das did her B.E. and M.Tech. in Computer Science. She received her Ph.D. from the School of Computer and Systems Sciences, Jawaharlal Nehru University, New Delhi, India in 2015. She is working as an Assistant Professor in the Computer Science Department at Assam University (a Central University), Assam, India since 2008. Her current research interest includes Mobile Ad-hoc Networks, Wireless Sensor Networks, Internet of Things, and Vehicular Ad-hoc Networks.

Dr. Nanhay Singh is working as a Professor in the Department of Computer Science and Engineering at Netaji Subhas University of Technology, East Campus, Delhi, India. Presently he is in the chair of Head of Department. His research areas are Web Engineering, Data Mining, Cloud Computing, IoT, and Ad-hoc Network. Dr. Singh has published many research articles in reputed international journals and conferences. Dr. Singh is also a member of various professional bodies and delivered many expert talks in reputed universities and institutes. He is supervising many research scholars. He has also organized various international conferences and faculty development programs on various latest research areas. He is also a member of selection committee of various organizations.

Contributors

N. Ambika
Department of Computer Applications
Sivananda Sarma Memorial RV College
Bangalore, India

Allam Balaram
Department of Information Technology
MLR Institute of Technology
Hyderabad, India

Vinayak Ashok Bharadi
Department of Information Technology
Engineering
Mumbai University, FAMT
Ratnagiri, India

P. Chandana
Department of CSE
Vignan Institute of Technology and
Science
Hyderabad, India

Rinki Chauhan
Computer Science and Engineering
Guru Gobind Singh Indraprastha
University
Delhi, India

Y. P. Chawla
Australian Graduate School of
Leadership
Sydney, Australia

Indrani Das
Department of Computer Science
Assam University
Silchar, India

Sanjoy Das
Indra Gandhi National Tribal University
Regional Campus, Manipur, India

Devraj
GGS Indraprastha University
New Delhi, India

Mehajabeen Fatima
Department of Electronics and
Communication
SIRT
Bhopal, India

Sanjeev Kumar Gupta
Department of Information Technology
Rabindra Nath Tagore University
Bhopal
India

Pravin S. Jangid
Computer Engineering Department
LRTCOE
Thane, India

Amanjot Kaur
IKGPTU
Jalandhar, India

Trupti S. Kedar
Mumbai University, FAMT
Ratnagiri, India

Munsifa Firdaus Khan
Department of Computer Science
Assam University
Silchar, India

Afreen Khursheed
Department of Electronics and
Communication
IIIT Bhopal
Bhopal, India

Nitima Malsa
Department of Computer Science
Banasthali Vidyapith
Banasthali, India

Mamta Meena
Computer Science and Engineering Department
Vikrant Institute of Technology and Management
Indore, India

Shaik Abdul Nabi
Department of CSE
Sreyas Institute of Engineering and Technology
Hyderabad, India

K. L. Neela
Department of Computer Science and Engineering
University College of Engineering
Thirukkuvalai, India

Jayashree Patil
School of Computer Engineering &Technology
MIT Academy of Engineering
Alandi(D) Pune, India

Sanjay S. Pawar
Department of ExTC, UMIT
SNDT Women's University
Mumbai, India

R. K. Ramesh
Department of Computer Science
Amirta Vidyalayam
Nagapattinam, India

Ram Shringar Rao
Department of Computer Science and Engineering, Netaji Subhas University of Technology
East Campus, Delhi, India

Rajesh Sachdeva
PG Department of Computer Science
Dev Samaj College for Women
Ferozepur, India

Gitimayee Sahu
Department of ExTC, UMIT
SNDT Women's University
Mumbai, India

Nandini Sidnal
Department of Computer Science & Engineering
KLEs M.S.Sheshagiri College of Engineering
Belgaum, India

Manda Silparaj
Department of CSE
ACE Engineering College
Hyderabad, India

Parminder Singh
Department of Information Technology
CGC Landran
Ajitgarh, India

Pooja Singh
Department of Computer Science and Engineering
Amity University, India

Suresha
Department of Computer Science & Engineering
Sri Venkateshwara College of Engineering
Bengaluru, India

Sitendra Tamrakar
Department of Computer Science and Engineering
Nalla Malla Reddy Engineering College
Hyderabad, India

Ramesh Vishwakarma
Department of Information Technology
Rabindra Nath Tagore University
  Bhopal
India

Vaibhav Vyas
Department of Computer Science
Banasthali Vidyapith
Banasthali, India

About This Book

Recent developments in wireless networks such as Mobile Ad-hoc Networks (MANETs), Vehicular Ad-hoc Networks (VANETs), and Wireless Sensor Networks (WSNs) open lots of opportunities for academicians and researchers to explore the area. The networking demand is increasing to a huge scale, and Big-data and cloud computing technologies help to manage the enormous volume of data. Nowadays, these data are placed on cloud called cloud data and described as Big-data. Researchers and data scientists using IoT include rapidly growing data that can be handled by using Big-data and its analytics. Big-data and cloud computing enable the enhancement of these networks. The VANETs help in managing vehicular traffic networks worldwide, optimize traffic management, improve traffic efficiency, etc. VANET has been developed for Intelligent Transportation Systems (ITS), and it is the future direction for the road transportation system. Current development in wireless communication, ad-hoc network, computing paradigms, Big-data, and cloud computing fetched the attention on the enhancement of intelligent devices equipped with wireless communication capabilities and high-performance processing tools. Therefore, the Big-data on the cloud will profoundly impact the implementation of ad-hoc networks such as VANETs, which make transportation systems much safer, efficient, and effective.

In recent developments, cloud computing, Big-data, and IoT-based services are used in various ad-hoc network applications such as MANETs, VANETs, FANETs, and WSNs to handle these networks better. Cloud computing is becoming an emerging research area across various wireless networks. With the help of cloud computing, the assessment, manipulation, sharing data, files, and programs to perform on-demand through the Internet. Cloud computing is considered as a key element to satisfy these requirements for various wireless networks. It is a network access model that aims to exchange a large number of computing resources and services. Integration of cloud computing and Big-data includes the various heterogeneous technologies that work together and provide real-time solutions for VANET and its applications.

Our target is to analyze various wireless ad-hoc networks and the role of cloud computing, IoT, and Big-data in managing the massive volume of data generated from the different wireless mediums. This book is ideal for academicians, researchers, engineers, and many other technical professionals. Further, this book will be very much useful and helpful to the UG/PG students and research scholars to understand the advancements such as edge computing, blockchain, routing, mobility models, and cloud architecture in cloud computing, Big-data, and IoT. Finally, these advancements improve the area of various wireless ad-hoc networks and its applications.

Chapter 1: *Cloud-Based Underwater Ad-hoc Communication: Advances, Challenges and Future Scopes*

Rinki Chauhan, Ram Shringar Rao, and Sanjoy Das

This chapter describes various issues, challenges, and future scope and the usage of cloud with the sensor nodes in an underwater ad-hoc communication network.

Chapter 2: *A Hybrid Cryptography Technique with Blockchain for Data Integrity and Confidentiality in Cloud Computing*
  K. L. Neela and R. K. Ramesh

  This chapter focuses on the construction of a secure cloud storage system in a decentralized way. Secure user authentication can be provided by using an iris verification system. Data confidentiality can be provided by using new hybrid cryptographic techniques such as Advanced Encryption Standard (AES), Data Encryption Standard (DES), and CST (Cyclic Shift Transposition). Besides, secure data retrieval and data integrity can also be provided by using data matrix code and blockchain verification scheme.

Chapter 3: *Fog Computing Environment in Flying Ad-hoc Networks: Concept, Framework, Challenges and Applications*
  Devraj, Ram Shringar Rao, and Sanjoy Das

  This chapter includes a thorough study of fog computing for FANETs and services based on fog computing. The issues and challenges for fog computing in FANETs are also included.

Chapter 4: *Wi-Fi Computing Network Empowers Wi-Fi Electrical Power Network*
  Y. P. Chawla

  This chapter shares knowledge about electrical power to understand the cross-disciplinary subject and improves with computing support also much needed for remote data centers. The matter involves leadership in innovation, need of dedication of those who are interested in a deep dive in the field of knowledge.

Chapter 5: Big Data Analytics for Vehicular Edge Networks
  Jayashree Patil, Suresha, and Nandini Sidnal

  This chapter focuses on the integration of edge computing with VANETs and performs data analytics. Edge computing technology has various benefits in vehicular networks in terms of efficient bandwidth, minimum latency, and context awareness. Also discussed are the survey on VANET, cloud computing, and essential concepts for edge computing in the transformation domain. Firstly, it explains the data gathering and utilization at edge nodes in different formats and identifying the most suitable format for storage on which various data mining techniques will classify the data. Secondly, data prefetching reduces data access time, stores data till analytics, and disseminates messages. Since edge nodes have lesser data storage capability, the data will be moved to cloud servers for future reference.

Chapter 6: *Impact of Various Parameters on Gauss Markov Mobility Model to Support QoS in MANET*
  Munsifa Firduas Khan and Indrani Das

  This chapter focuses on the Quality of Service (QoS) issues of MANET. To study QoS's effect, many experiments were conducted on Ad-hoc On Demand Distance Vector (AODV) routing protocol with the Gauss Markov mobility model with other network parameters.

Chapter 7: *Heterogeneous Ad-hoc Network Management: An Overview*
Mehajabeen Fatima and Afreen Khursheed

This chapter reviews different types of networks: heterogeneous ad-hoc network (HANET), basic concepts, common features, issues, and research trends. Although various studies are accessible on the topic, rapid developments require a refreshed record on the topic. In the first segment, history, wired and wireless network design approach, enabling technologies are discussed. The second segment of this chapter presents an outline of different arising models of HANET and discusses their particular properties. A brief detail is carried out on versatile ad-hoc networks such as MANET, VANET, Wireless Mesh Network (WMN), and WSN. Further segments investigate the basic features and issues of HANET. The chapter closes by presenting a summary of the intelligent management requirements of HANET and directions for further research.

Chapter 8: *Deployment of the Biometrics-as-a-Service (BaaS) Design for the Internet of Biometric Things (IoBT) on the AWS Cloud*
Vinayak Ashok Bharadi, Trupti S. Kedar, Pravin S. Jangid, and Mamta Meena

This chapter discusses compact sensors and biometric capture gadgets; biometric verification will become the Internet of Biometric Things (IoBT). Widespread use of biometric authentication has resulted in many authentication requests, and cloud-based execution is a practical alternative for handling such a large number of requests. The advancement of a multimodal biometric validation framework is examined in this chapter.

Chapter 9: A Comprehensive Survey of Geographical Routing in Multi-hop Wireless Networks
Allam Balaram, Manda Silparaj, Shaik Abdul Nabi, and P. Chandana

This chapter surveys geographical routing and its mobility issues in prominent Multi-hop Wireless Networks (MWNs) such as MANET, FANET, WSNs, and VANET.

Chapter 10: *Energy-Aware Secure Routing in Sensor Network*
N. Ambika

This chapter focuses on energy-aware secure routing in sensor networks. To save energy, the proposed method minimizes the load on the static nodes and increases their lifespan.

Chapter 11: *Deploying Trust-Based E-Healthcare System Using Blockchain-IoT in Wireless Networks*
Amanjot Kaur and Parminder Singh

This chapter discusses about linking Blockchain to IoT and looks at its various expressions. How IoT and blockchain technologies are used for e-healthcare is discussed. IoT devices are used to monitor health care, which involves collecting and evaluating patient data periodically. The data collected is kept in a secure block by the blockchain, and the patient will be directly connected to healthcare professionals at hospitals or healthcare center.

***Chapter 12**:* Low Cost Robust Service Overloading Fusion Model for Cloud Environments
Sitendra Tamrakar, Ramesh Vishwakarma and Sanjeev Kumar Gupta

In this chapter, a new scheme comprised of service exchange has been executed to give the intended security in the environment and lessen security risks to the cloud consumers. It builds the Service Overloaded Cloud Server (OCS) and provides a secure cloud environment. The service overloading technique will offer speedy and protected entree to cloud services underneath the cloud security atmosphere even once users are in motion. This authentication method doesn't need to hold any device perpetually compared to the traditional service overloading technique. There's no value for getting or buying any hardware and might offer a quick entree to the cloud service with a suitable authentication method.

Chapter 13: *Load Balancing Based on Estimated Finish Time of Services*
Rajesh Sachdeva

In this chapter, a load-balancing procedure that tries to advance the system routine based on the estimated finish of service time has been presented. The technique discussed improves the system routine by improving its waiting time, response time, and processing time.

Chapter 14: *Blockchain-Enabled Smart Contract Optimization for Healthcare Monitoring Systems*
Nitima Malsa, Vaibhav Vyas, and Pooja Singh

In this chapter, a smart contract is proposed for the healthcare system. Ethereum platform has been chosen for developing smart contracts

Chapter 15: *Interference Mitigation Using Cognitive Femtocell from 5G Perspective*
Gitimayee Sahu and Sanjay S. Pawar

This chapter discusses about cognitive femtocell which is an efficient solution for proper resource management and supports the guaranteed QoS of the user.

1 Cloud-Based Underwater Ad-hoc Communication
Advances, Challenges, and Future Scopes

Rinki Chauhan
Guru Gobind Singh Indraprastha University

Ram Shringar Rao
Department of Computer Science and Engineering
Netaji Subhas University of Technology

Sanjoy Das
Department of Computer Science
Indra Gandhi National Tribal University-
Regional Campus Manipur

CONTENTS

1.1 Introduction 2
1.2 Communication with the Sensors 3
1.3 Connecting the Sensors with the Cloud 4
    1.3.1 Architecture of the Underwater Sensor Network with Cloud Computing 5
1.4 Various Outcomes of Cloud Integration 7
    1.4.1 A Trust Model Based on Cloud Theory in Underwater Acoustic Sensor Networks 7
    1.4.2 An Energy-Balanced Trust Cloud Migration Scheme for Underwater Acoustic Sensor Networks 7
    1.4.3 Bidirectional Prediction-Based Underwater Data Collection Protocol for End-Edge-Cloud Orchestrated System 8
    1.4.4 An Underwater IoT System, Creating a Smart Ocean 8
    1.4.5 CUWSN: An Energy-Efficient Routing Protocol for the Cloud-Based Underwater Ad-hoc Communication Network 9
    1.4.6 SoftWater: Software-Defined Networking 9
1.5 Various Challenges in a Cloud-Based Underwater Communication Network 10
1.6 Future Scope 11
References 12

DOI: 10.1201/9781003206453-1

1.1 INTRODUCTION

Recently, underwater expeditions have increased with the use of technology. The human can only survive up to a certain limit of underwater pressure. Therefore, researchers have started working on different devices, sensors that can withstand that pressure and can be used to collect data on behalf of humans. The past methods include designing such a vehicle with certain features equipped in it, and if more features are required, the sensors are attached to it accordingly. However, this methodology has caused a waste of resources; therefore, a new and improved architecture was proposed, known as the Ad-hoc network, which causes different devices or sensors to create a quick connection with each other and communicate through that network [1–5].

With this advancement, the single vehicle is now dismounted into various nodes. Each node performs a separate task and communicates and coordinates with each other through the Ad-hoc network, as shown in Figure 1.1. This architecture is known as Underwater-based Wireless Ad-hoc Network (UWAN).

As explained in Figure 1.1, an underwater Ad-hoc network involves a sensor network with various sensor nodes which are placed underwater, anchored or moving, along with different Autonomous Underwater Vehicles (AUVs) and sink (receiver) nodes used for gathering and relaying the captured data to the respective sink nodes to be used for various underwater applications such as water pollution monitoring, oceanographic data collection, disaster prevention, and applications involving navigation and tactical surveillance [6].

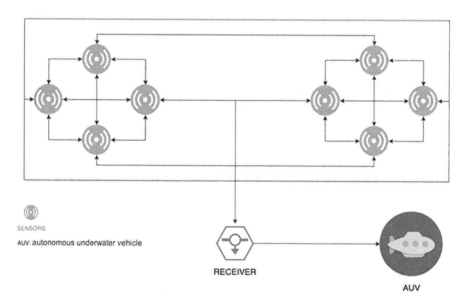

FIGURE 1.1 An underwater Ad-hoc network [6].

1.2 COMMUNICATION WITH THE SENSORS

With the sensors deployed, the next hurdle is communication, as getting outputs from underwater sensor networks can be very challenging due to varying currents, changing marine environment, multipath propagation, propagation delay, high error rates, limited bandwidth, limited battery life of sensor node, high attenuation, etc. In order to prevent the issues mentioned above, sonic transducers were being used to transmit signals, but now a wireless transmission system has been introduced, and using motes over these sonic transmitters is a more cost-effective method. This wireless communication method provides two big advantages: reduced latency, as instead of sound, electromagnetic waves are being used; second, these electromagnetic waves can work on a higher frequency, reducing the risk of interference [7].

Therefore, communication can be done by four methods:

a. Radio Frequency (RF) Communication:

An RF signal is a wireless electromagnetic signal. It has been highly appreciated and used in free space by military and civilians as well. However, certain characteristics cause high attenuation in RF signals in water.
   i. Electromagnetic properties of water, as permittivity and conduction are both higher in water than air, the phase velocity is comparatively lesser in water than in air.
   ii. Transition frequency, as attenuation is directly proportional to transition frequency, lower transition frequency is preferred.

Apart from the above properties, RF communication also depends upon the antennas being used, as the wavelengths are reduced inside water, the antenna size for this communication also needs to be small. When the signal does get to cross the surface of water to enter free space, only a part of it is able to pass the surface of the water, due to total reflection; not only that, these waves are further refracted and they appear to have propagated along the surface of the water [8].

b. Magneto-Inductive (MI) Communication:

MI communication is the result of using coils which are very tightly tuned with quasi-static field. The outcome is a magnetic field, which is produced after a time-varying current goes through a coil. Many factors based on the coil affect this magnetic field, such as size, shape, and magnitude of current. However, the main factors that affect MI communication are permeability and conductivity of the medium being used. The permeability is similar to 1, which is good enough for MI to produce a robust communication architecture for water usage. For the case of conductivity, water in oceans is salty, causing high conductivity; hence, it will affect communication as well. Mainly, MI is being used in two scenarios: bridge scour monitoring and levee scour monitoring. MI is being used in many underground structural monitoring applications as well apart from underwater as well.

The highest efficiency of MI communication has been encountered in the freshwater experiment, in which the testing was based on the range between the source and the destination; the result indicates that the range reached hundreds of meters without any propagation delay or fading due to its superior bit error rate performance [9].

c. Optical Communication:

Underwater Wireless Optical Communication (UWOC) is the method of sharing information from the underwater sensors to the receivers in the free space using light as a propagative method. It is the answer to the problem faced by AUVs and other underwater vehicles that they are not able to send their real-time feed to the receivers in the free space or allow users to tele-operate the vehicles due to limited bandwidth, high latency, and high transmission losses; however, UWOC is able to provide higher bandwidth usage, higher data rates transmission with lower latency rates at a lower cost, with higher lifetime.

Normally, water absorbs optical signals, and due to the particles present in the sea, there is high chance of optical scattering, but in the case of the visible spectrum, seawater does not absorb in the blue/green zone, which is where Underwater optical communication (UWOC) is exploiting this characteristic to provide a more robust communication architecture, and with seawater, high-speed connection is created. However, when a water body with suspended matter is being used for UWOC implementation, there will be slight delays in the delivery of the packets, and there is high possibility of interference, causing the packets to replace themselves. The performance is also based on the amount of turbulence present in the water body, which is basically derived from the temperature and amount of salt in the water body [10,11].

d. Acoustic Communication:

Wherever there are particles, there are ways for sound to travel. Therefore, for communicating underwater, the sound signals created by the hydrophones are used in acoustic communication architecture. Since the sound particles are not absorbed by the water, they are preferred the most for a propagation method. That is why they allow longer travel rates, but at a rate much slower than other methods [12].

Each method is having its own merits and demerits under different environment and different priorities. Out of the above methods, acoustic method of communication is the most popular in underwater communication. Acoustic communication is good as it is having long range of communication, but there are various issues in that also such as limited bandwidth, multipath and fading among the channel, propagation delay, high data error rates, and limited battery power.

1.3 CONNECTING THE SENSORS WITH THE CLOUD

The main purpose of having these sensors in the water bodies is to collect data, record various properties, and capture different scenarios that are different for different water bodies. But the main purpose of collection of such data is to analyze it

Cloud-Based Underwater Ad-hoc Communication

and create conclusions and predictions for future. Analysis of stored data can result in understanding the current situation properly and can guide us to make some future predictions based on analysis of the data stored. Also, it would help the corresponding users to take prompt action in a timely manner based on their observations and analysis. The most efficient and fast way to share this data to every lab is to centralize it and provide authenticated access. Therefore, after gathering of such data, Cloud storage can store the successfully transmitted data from the underwater Ad-hoc network and distribute this data to the customer as per their filtered requirement after proper authorization, thereby making it available for different types of users as per their need; it would also provide ease of access along with various access privileges from anywhere and anytime [13,14]. This way we can ensure the data is available to users with proper security mechanisms which ensures data integrity and consistency in data storage and usage. Not only this, the data can be stored for several decades without needing to worry about the storage component as the Cloud can sustain large amount of data and its processing [15–17].

1.3.1 Architecture of the Underwater Sensor Network with Cloud Computing

The pioneers of the field shared a draft of the architecture to be used while using Cloud with the underwater Ad-hoc network. This architecture is heavily based on visualization, which helps creating a complete abstraction of the underlying physical structure of the Ad-hoc network, as mentioned in Figure 1.3. It uses the same method of abstraction as of sensor nodes; thus, the underwater sensor network integrated with Cloud is independent of topology or the proximity requirements of the different parts of the architecture [18].

The architecture mainly consists of three layers as shown in Figure 1.2.

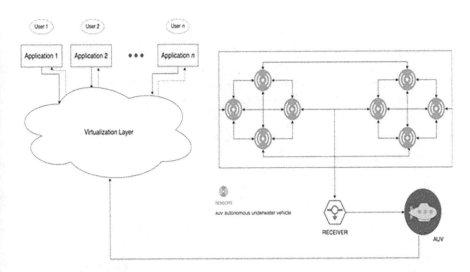

FIGURE 1.2 Architecture of the underwater sensor network with Cloud computing [18].

a. Application Layer

This layer consists of the actual users, which could be either the owner or the employees who are supposed to run analytical programs on the data collected by the sensors over the Cloud. With the help of Cloud, the applications can be used by anyone on the planet, but with proper authentication.

b. Virtualization Layer

The virtualization layer is the layer of Cloud computing. This layer provides services to the users in the application layer. It contains a certain number of virtual groups, which actually perform the requests created by the users. It can be said that these virtual groups are the interpreters of the Cloud architecture.

c. Underwater Network Layer

The underwater network layer is the collection of all the sensors in the network being used. It comprises two major devices, the sensor nodes and the AUV. The sensor nodes collect the data, which is then received by the AUVs to form a proper package and then transmit the data to the topological receivers.

The sensor used in an underwater Ad-hoc communication network mainly comprises three units: sensing unit, processing unit, and communication unit. The architecture of the sensor nodes is shown in Figure 1.3.

a. Sensing Unit

Sensing unit comprises the electrical components which record the data. The data being stored is in analog format, which needs to be converted into a digital signal to be used for signal processing via computers. Therefore, an ADC (analog to digital converter) is attached. The data is transferred to the next part for storage and processing purposes.

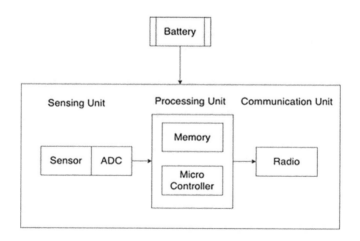

ADC: Analog to Digital Converter

FIGURE 1.3 Architecture of sensor node [19].

b. Processing Unit

In this unit, the digital outcome of the previous unit is temporarily stored for processing purposes, with which noise and unnecessary signals are removed. Cloud computing also performs the same operations, but that is a new technology; before Cloud computing, this task was conducted in the sensor itself. However, on Cloud, removal of noise is much more powerful.

c. Communication Unit

The communication unit is only made up of a radio component, which actually transmits the data to the AUVs. The techniques for transmission have already been studied in Section 1.2.

1.4 VARIOUS OUTCOMES OF CLOUD INTEGRATION

There have been many advancements in underwater Ad-hoc communication networks and its integration with Cloud [3]. With the help of Cloud, the whole process can be more efficiently performed and even use less power. With the help of Cloud, machine learning can also be implemented in the analysis and research based on the data, which has performed various tasks that humans are incapable of, such as predicting future data sets based on the trends found in the current data sets. Such advancements are mentioned below.

1.4.1 A Trust Model Based on Cloud Theory in Underwater Acoustic Sensor Networks

With the system being online, the underwater sensor networks are now vulnerable to large number of attacks, and since these sensors are not using the traditional architecture for communication as the free space sensors on terrestrial land, the security requirements are different as well. Therefore, a new model, Trust Model, for security mechanisms is introduced in the industry, which is specifically designed for sensors used in a free environment. The major role of using this trust model is to ensure that fuzziness and uncertainty are removed and evaluation accuracy is improved. This model also detects malicious nodes in the network and verifies successful packet delivery.

This model is based on the Cloud theory, Trust model Cloud (TMC), which requires the generation of trust evidence between the nodes/sensors, for that, trust calculation is necessary, which can be done in two different ways, either to be recommended by a node already present in the system, or calculate trust directly. In trust evidence calculation, the packets are being monitored, as well as the transmission outcomes, such as packet loss, error rate, and level of energy consumption. In the event of an attack, these values change the usual pattern [19].

1.4.2 An Energy-Balanced Trust Cloud Migration Scheme for Underwater Acoustic Sensor Networks

With advancements in the security management of the nodes, it is necessary to make sure that these advancements are not hampering the performance of the sensors in any way possible. However, in the trust model suggested for security management of the underwater Ad-hoc networks, there is an issue of power management, as the trust evidence are required for verification and evaluation of the packets; the trust evidence are collected so frequently that these nodes consume their power well before their expected expiry and are dying prematurely. This happens majorly in those nodes which are closer to the trust Cloud node. Hence, a migration scheme was proposed to make sure that the issues above can be mitigated. This migration scheme consists of calculations for determining the destination node, transfer of packets, and update mechanism for logging purposes.

Since Cloud allows a cluster-based architecture, the destination node cluster is calculated in the initial phase, based on the residual energy of the sensors. A cluster ability (CA) technique is used to find the actual destination node inside the destination cluster. The destination node is considered as the new Trust Cloud node. Every time the packet transfer is made, the Cloud node data is transferred to a new location, as the calculation for the destination cluster is a hierarchical process. This methodology balances energy consumption and prevents premature death of sensors in the underwater Ad-hoc network [20].

Machine learning, the power of computing to perform tasks incapable by human hands, is being used to improve technology at an incredible rate. With its application in the underwater Ad-hoc network architecture, analytical research has found that the performance of the underwater nodes can be increased with more efficient power management [21].

1.4.3 BIDIRECTIONAL PREDICTION-BASED UNDERWATER DATA COLLECTION PROTOCOL FOR END-EDGE-CLOUD ORCHESTRATED SYSTEM

The data from the sensors is collected for analytical purposes. With the data now being on Cloud, the next step in this field is edge-cloud computing, which is basically bringing the computing cluster closer to the location where it is actually needed. In the traditional ways of communication from underwater sensors, the data collection rate is very slow and inefficient in many ways. The better alternative is to predict the data. The outcomes of predicting data will be reduced energy consumption and relieved communication bandwidth. Having the edge computing prospect will lead to faster responses from the server as it will be closer to the actual clusters of sensors/nodes.

The AUVs can act as the edge elements to provide the necessary services to the edge computing clusters. The task, however, is to verify the predictions, whether they are correct or not; therefore, a bidirectional system is to be used in this scenario, where the edge computing cluster will share its predictions with the sensors underwater and get a confirmation from them, using the bidirectional tunnel, as these sensors, during the time taken by edge computing clusters to predict the data, have already collected the data. If the predicted data matches the collected data, there will not be any need to receive the same data; however, if it doesn't, the actual data will be transmitted from the sensors [22].

1.4.4 AN UNDERWATER IoT SYSTEM, CREATING A SMART OCEAN

With the improvement in technology, the researchers have started working on a smart ocean architecture that is based on the idea that different highlights of the ocean should be investigated and perceived. The Underwater Internet of Things (UIoT), an augmentation of the IoT to the submerged climate, establishes incredible innovation for accomplishing the concept of smart ocean. The UIoT became a possibility with increment in the advances being made in the framework designs, applications, mitigating difficulties, and open issues in the underwater Ad-hoc communication network. Apart from the underwater architecture, the Cloud and the coastal clusters are being used to create the UIoT. The data fusion techniques used by the clusters are more powerful than the Cloud; however, to provide universal access, the data is then transferred from these coastal clusters to the Cloud. The data can be accessed by anyone, with proper authentication.

The Cloud is able to use deep learning effectively and uses a parallel system for faster processing; hence, most of the processing and analysis work is done on the Cloud. The coastal clusters then use the results to make intelligent decisions [23].

1.4.5 CUWSN: An Energy-Efficient Routing Protocol for the Cloud-Based Underwater Ad-hoc Communication Network

There are many different methods of communicating the data from the sensor to the Cloud. However, the methods do not support and control of data while transmission. However, Cluster-based Underwater Wireless Sensor Network (CUWSN) is made to provide support and control of the scenarios in the underwater medium while keeping the transmission as energy efficient as possible to maintain the lifespan of the sensor nodes. The architecture of this routing algorithm is based on the cluster head technology, which increases throughput, low packet loss, low transmission delay, and higher packet delivery while providing control to the users. The life of the equipment in use is increased with the help of the multi-hop technique of transmission. However, limitations of this protocol include a short range of communication, smaller coverage of the area, and mediocre security [24].

1.4.6 SoftWater: Software-Defined Networking

There are many different architectures present in the industry. However, they are all based on the same architecture with a few tweaks made to meet the required specifications. The traditional architecture is completely hardware-based, which makes it inflexible; thus, adopting new technologies causes a lot of trouble for the architects to design a better system. Therefore, a software-based architecture was proposed, which is the next-generation paradigm. This method helps by simplifying the network, improves efficient utilization of the system, makes the network highly flexible, and promotes innovation. The first software in the industry to perform these tasks is known as "SoftWater." This software provides the following functionality:

  i. Add new nodes to the system/network easily
  ii. Enable interoperability
  iii. Truly differentiated and scalable network

SoftWater has an inherent feature of "Network Function Virtualization," which takes hardware functionality from the hardware, and performs the same function in the Cloud. Thus making it highly flexible, which comes with the advantage of quick adaption to changes made by the user, it also controls the globally optimized network created by an organization, which owns the sensor nodes and AUVs. SoftWater increases the throughput by implementing three-layer functions in a single layer of virtualization, namely, network, Media Access Control (MAC), and physical layers [25].

1.5 VARIOUS CHALLENGES IN A CLOUD-BASED UNDERWATER COMMUNICATION NETWORK

Most of the challenges faced in an underwater Ad-hoc communication network rises from water itself, as it provides the highest permittivity causing packet loss at a higher rate. Moving forward from the technical difficulties, oceans are the most undiscovered part of the planet, and therefore the undiscovered part's chemical and physical properties are still a mystery to humans.

For any underwater Ad-hoc network, the requirement is of a distributed, self-organizing structure that can reorganize itself with the varying environment and lasts longer with proper relay and sleep modes of the participating sensor nodes. The sensor nodes' life is crucial since we cannot recharge them as we do in terrestrial Ad-hoc networks. Also, the replacement of these nodes will add a great burden of cost. Thus, in every underwater Ad-hoc communication, we need to emphasize on saving the battery of participating sensor nodes by focusing on relay mechanisms used in the network. Various routing algorithms used in the different network also helps to choose the data transfer by multi-hop or by single-hop and which node to take or share the burden of data relay at what time intervals. All this helps to minimize the dissipation of energy of the sensor nodes [26,27].

Broadly, two conditions need to be fulfilled for mitigating the issues mentioned above; such conditions are based on the difficulties faced during the setup of the UWAN.

   a. Which Communication Channel Is Right?

   Different communication channels have already been discussed in Section 1.2; most of the communication channels are well suited for mediums such as air. However, it is difficult to choose for water as it does not allow much of the transmitted data ever to leave the medium. A comparison is presented in Table 1.1 of all the different channels mentioned in

TABLE 1.1
Comparison of Different Channels

Metrics	Radio Waves	Acoustic Waves	Optical Waves
Latency	Moderate	Highest	Lowest
Speed	Moderate	Lowest	Highest
Mobility	Lowest	Moderate	Highest
Attenuation	Yes	Yes	Yes
Frequency	Moderate	Lowest	Highest

TABLE 1.2
Different Routing Protocols

Protocol	Characteristics	Limitations
Near infra-red	Highly efficient, greedy approach using path selection.	Works on only a single path.
Rebar	Highly energy efficient, based on geographical routing.	Efficiency is decreased when in contact with an obstacle.
Low-energy adaptive clustering hierarchy (LEACH)-based algorithm	Provides multi-hop packet delivery, works on cluster-based technology.	Range is limited.
Energy Aware Routing Protocol	Highly energy efficient, based on geographical routing.	Implementation cost is high.
SEANAR	Routing is based on quality.	Complexity is very high.
Mobile geo-cast routing protocol	Provides support for 3-D mapping of data underwater.	Power consumption increases when water current speed is high.

Section 1.2, which can help suitable readers to make conclusions for which channel to be used in an UWAN [23].

b. Which Routing Protocol to Choose?

A comparison of routing protocols is shown in Table 1.2. However, they need to be restructured for the sake of underwater communication to be feasible. Therefore, some of the protocols that allow modification in their algorithm are mentioned below with their characteristics and limitations,

The correct outcome of an application based on an UWAN requires rigorous testing of both the communication channel and the routing algorithm before its implementation. For the purpose of testing, simulations can be made. Energy efficiency is of the utmost importance while building an UWAN architecture, as the sensor nodes need to be recharged with fuel, they need to be brought of the water, and that is not a desirable characteristic. Therefore a routing algorithm with energy efficiency should be prioritized. Also, these devices are expensive, and removing them continuously from their location can cause physical damage to them as well. Water bodies do not allow the property of full duplex systems to operate, as the sizes of ocean bodies are very large, and the transducers used for sending and receiving the signals cannot be separated for large distances. Hence they can only perform half-duplex operations. Also, the size of the transducers depends on the size of the water bodies and the wavelength of the frequency being used in the communication architecture [28].

1.6 FUTURE SCOPE

Advancement in the field of UWAN can benefit many older and newer applications, majorly the military. A framework built on UWAN would have the option to take care of job all the more productively in the vicinity of distant/remote spots, where it's less achievable to do information gathering safely. It could likewise diminish the dangers associated with the submerged military application. Moreover still, endeavors must be made to expand the research in the area with the goal that a completely solid framework can be sent and utilized proficiently. New better software and simulation devices must be built with the goal that the exact model can be actualized. A legitimate portability model must be created or adjusted, thinking about water as a correspondence medium. For better understanding and clarification, the model should have the option to analyze and build upon the impact of different ongoing elements like the actual property of channel, multiplexing and modulation techniques, working frequency on submerged routing protocols.

REFERENCES

1. Raw, R.S., & Lobiyal, D.K. B-MFR routing protocol for vehicular ad hoc networks, *2010 International Conference on Networking and Information Technology*, Manila, Philippines, 2010, pp. 420–423.
2. Raw, R.S., & Das, S. (2013) Performance analysis of P-GEDIR protocol for vehicular ad hoc network in urban traffic environments. *Wireless Personal Communications 68*, 65–78.
3. Rana, K.K., Tripathi, S. & Raw, R.S. (2020). Link reliability-based multi-hop directional location routing in vehicular ad hoc network. *Peer-to-Peer Networking and Applications 13*, 1656–1671.
4. Rana, K.K., Tripathi, S. & Raw, R.S. (2020). Inter-vehicle distance-based location aware multi-hop routing in vehicular ad-hoc network. *Journal of Ambient Intelligence and Humanized Computing 11*, 5721–5733.
5. Rana, K.K., Tripathi S., & Raw, R.S. (2017). Analysis of expected progress distance in vehicular ad-hoc network using greedy forwarding. In *11th INDIACom 4th IEEE International Conference on Computing for Sustainable Global Development* (pp. 5171–517), IEEE: Delhi.
6. Pranitha, B., & Anjaneyulu, L. (2020). Analysis of underwater acoustic communication system using equalization technique for ISI reduction. *Procedia Computer Science, 167*, 1128–1138.
7. Basagni, S., Conti, M., Giordano, S., & Stojmenovic, I. (Eds.). (2013). *Mobile Ad Hoc Networking: Cutting Edge Directions* (Vol. 35). John Wiley & Sons.
8. Stefanov, A., & Stojanovic, M. (2011). Design and performance analysis of underwater acoustic networks. *IEEE Journal on Selected Areas in Communications, 29*(10), 2012–2021.
9. Ahmed, N. (2017). *Magneto Inductive Communication System for Underwater Wireless Sensor Networks*. Missouri University of Science and Technology: Rolla, MI.
10. Schirripa Spagnolo, G., Cozzella, L., & Leccese, F. (2020). Underwater optical wireless communications: overview. *Sensors, 20*(8), 2261.
11. Saeed, N., Celik, A., Al-Naffouri, T.Y., & Alouini, M.S. (2019). Underwater optical wireless communications, networking, and localization: a survey. *Ad Hoc Networks, 94*, 101935.
12. Yang, T.C. (2012). Properties of underwater acoustic communication channels in shallow water. *The Journal of the Acoustical Society of America, 131*(1), 129–145.
13. Misra, S., Chatterjee, S., & Obaidat, M.S. (2014). On theoretical modeling of sensor cloud: a paradigm shift from wireless sensor network. *IEEE Systems Journal, 11*(2), 1084–1093.

14. Kumar, S., & Raw, R.S. (2018) Flying ad-hoc networks (FANETs): current state, challenges and potentials. *5th International Conference on "Computing for Sustainable Global Development", Bharati Vidyapeeth's Institute of Computer Applications and Management (BVICAM)*, New Delhi, India, IEEE Conference ID: 42835, (pp. 4233–4238).
15. Gaba, P., & Raw, R.S. "Vehicular Cloud and Fog Computing Architecture, Applications, Services, and Challenges." In *IoT and Cloud Computing Advancements in Vehicular Ad-Hoc Networks*, edited by Ram S.R., Jain, V. Kaiwartya, O. & Singh, N. 268–296. IGI Global: Hershey, PA.
16. Aliyu, A., Abdullah, A.H., Kaiwartya, O., Cao, Y., Usman, M.J., Kumar, S., Lobiyal, D.K. & Raw, R.S. (2018) Cloud computing in VANETs: architecture, taxonomy, and challenges. *IETE Technical Review*, *35*(5), 523–547.
17. Kumar, M., Yadav, A.K., & Khatri, P. (2018) Global host allocation policy for virtual machine in cloud computing. *International Journal of Information Technology*, *10*, 279–287.
18. Maraiya, K., Kant, K., & Gupta, N. (2011). Study of data fusion in wireless sensor network. In *International Conference on Advanced Computing and Communication Technologies* (pp. 535–539), San Diego, CA.
19. Jiang, J., Han, G., Shu, L., Chan, S., & Wang, K. (2015). A trust model based on cloud theory in underwater acoustic sensor networks. *IEEE Transactions on Industrial Informatics*, *13*(1), 342–350.
20. Han, G., Du, J., Lin, C., Wu, H., & Guizani, M. (2019). An energy-balanced trust cloud migration scheme for underwater acoustic sensor networks. *IEEE Transactions on Wireless Communications*, *19*(3), 1636–1649.
21. Pramod, H.B., & Kumar, R. (2016, October). On-cloud deep learning based approach for performance enhancement in underwater acoustic communications. In *2016 International Conference on Signal Processing, Communication, Power and Embedded System (SCOPES)* (pp. 621–624). Odisha: IEEE.
22. Wang, T., Zhao, D., Cai, S., Jia, W., & Liu, A. (2019). Bidirectional prediction-based underwater data collection protocol for end-edge-cloud orchestrated system. *IEEE Transactions on Industrial Informatics*, *16*(7), 4791–4799.
23. Qiu, T., Zhao, Z., Zhang, T., Chen, C., & Chen, C. P. (2019). Underwater Internet of Things in smart ocean: system architecture and open issues. *IEEE Transactions on Industrial Informatics*, *16*(7), 4297–4307.
24. Bhattacharjya, K., Alam, S., & De, D. (2019). CUWSN: energy efficient routing protocol selection for cluster based underwater wireless sensor network. *Microsystem Technologies*, 1–17.
25. Akyildiz, I.F., Wang, P., & Lin, S.C. (2016). SoftWater: software-defined networking for next-generation underwater communication systems. *Ad Hoc Networks*, *46*, 1–11.
26. Rana, K.K., Tripathi, S., & Raw, R.S. (2016). Analysis of expected hop counts and distance in VANETs. *International Journal of Electronics, Electrical and Computational System*, *5*(4), 66–71.
27. Conti, M., & Giordano, S. (2014). Mobile ad hoc networking: milestones, challenges, and new research directions. *IEEE Communications Magazine*, *52*(1), 85–96.
28. Singh, S.K., & Tagore, N.K. (2019, March). Underwater based adhoc networks: a brief survey to its challenges, feasibility and issues. In *2019 2nd International Conference on Signal Processing and Communication (ICSPC)* (pp. 20–25). Hualien: IEEE.

2 A Hybrid Cryptography Technique with Blockchain for Data Integrity and Confidentiality in Cloud Computing

K.L. Neela
University College of Engineering, Thirukkuvalai

R.K. Ramesh
Amirta Vidyalayam, Nagapattinam

CONTENTS

2.1 Introduction 16
    2.1.1 Security Issues in Cloud Computing 16
2.2 Related Work 17
2.3 Problem Definition...18
2.4 Objectives 19
2.5 Proposed Methodology...19
    2.5.1 Registration Phase 20
    2.5.2 Data Confidentiality Using Hybrid Algorithm 20
    2.5.3 Secure Data Integrity and the Transaction 24
        2.5.3.1 The Setup Phase..24
        2.5.3.2 Check Proof Phase 25
2.6 Performance Analysis..25
    2.6.1 The Simulation Results..26
    2.6.2 Signature Verification 27
2.7 Conclusion 27
References..27

2.1 INTRODUCTION

A mainframe computer is mainly used to handle a huge amount of data. Several end users can access the data in a centralized manner. Mainframe computer systems are expensive for companies to buy, even though it provides excellent throughput of I/O processing. After the usage of mainframe systems, many new developments have been raised in computing for better storage and sharing of resources. Mainly, cloud computing which is based on utility computing provides metered services for the resources, i.e., the customer has to pay for used resources. Basically, cloud computing is derived from a mainframe computer.

According to researchers, organizations, and cloud users, the cloud definition varies in some circumstances. The definition of cloud computing is to provide services such as platform, software, and hardware over the Internet. Cloud environment offers these services through on-demand and pay-as-you-go basis [1]. Centralized Cloud Service Provider (CSP) controls the cloud environment. The centralized system is untrustworthy because if hackers/unauthorized users steal the data, confidentiality will be compromised. In spite of security issues occurring in cloud environment, there are many advantages over cloud computing such as scalability and efficiency [2–4]. Cloud computing does not need any capital expenditure/investment, i.e., it allows rental services to use hardware and software. Sometimes, due to corruption of the hard disk, the data stored in the personal computer may be lost, whereas data stored in cloud computing environment can be accessed at any time and anywhere later on over the Internet. Cloud computing is mainly used in government, companies such as public and private concerns, universities, and research organizations.

Cloud computing was built to convey computing services over the Internet [5]. The fully developed ready-to-use applications and hardware resources like network storage are provided according to user needs [3,6]. Cloud computing is a famous and successful business plan because of its attractive qualities and features. However, the main difficulty of cloud storage is the integrity and confidentiality of information. This is because the cloud user is unaware of the locality of the data in the cloud environment, so an unauthorized user can easily edit the data without the knowledge of the legitimate user [7,8]. This makes the development of a secure storage system challenging due to unauthorized data access. In organizations, cloud computing is a centralized system that is very commonly used to store financial data, medical, military, education, personal data of employee, organization asset data, collaboration data, etc. The theft of the information can prompt awful complications for organizations. Several researchers use cryptographic techniques to overcome security issues [9,10].

2.1.1 Security Issues in Cloud Computing

Many advantages exist in cloud computing mentioned previously, but the major issues are security and integrity. Centralized service providers who are semi-trusted control the cloud environment so that hackers easily attack the CSP without the data owner's knowledge. Cryptography provides security by encryption and decryption process so that no one is authorized to reveal the secret information except the sender and recipient. Therefore, no third party can alter or reveal the data [11].

In recent years, numerous cryptographic strategies have been utilized. Some of the cryptographic algorithms like DES (Data Encryption Standard), 3DES, AES (Advanced Encryption Standard), RSA (Rivest–Shamir–Adleman), RC4 (Rivest Cipher 4), and Blowfish [12,13] were proposed to secure the information. However, the existing cryptographic algorithms provide less efficiency and security and take more time for the encryption process and decryption process of data. To overwhelm these issues, a new hybrid cryptographic technique that includes AES, DES, and CST (Cyclic Shift Transposition) algorithms is proposed. The proposed system concentrates on the construction of a secure cloud storage system in a decentralized way. The proposed work does not depend on the third-party system. In this scheme, secure user authentication can be provided using an iris verification system and data confidentiality can be provided by using a hybrid algorithm. In addition, secure data integrity can be provided by using blockchain verification mechanism and data matrix code. The proposed system provides more security and integrity in the cloud environment in a well-organized manner. In this scheme, data ownership is secured. The service provider cannot access the data without the knowledge of the data owner.

2.2 RELATED WORK

Goyal and Kant [14] proposed a hybrid cryptography algorithm in cloud computing to achieve cloud data protection. The proposed hybrid algorithm comprising a different cryptography approach includes SHA-1 (Secure Hashing Technique), AES, and ECC (Elliptic Curve Cryptography). Data are shared from the owner to the cloud to the end cloud user. As the cloud combines dissimilar resources to offer the services, many vulnerabilities may exist in a cloud setup, whose exploration may be horrific for cloud data storage. It performs encryption as a primary security policy. Encoding is the technique for altering plaintext information in an encrypted form of ciphertext that can be deciphered and read by the legitimate person having a valid decryption key only. The illegitimate or mischievous person cannot easily decrypt and interpret the ciphertext without the decryption key. The proposed method sends two protected keys for high security. Although, it reduces the speed of the data while using One Time Password (OTP).

Lee et al. [15] implemented AES in Heroku cloud platform for data security. The steps involved in this AES algorithm are substitute bytes, shift rows, mix columns, and ADD round key. Here, heroku is a cloud platform that supports programming language. Dynop app is considered as the heart of the cloud. Before uploading the data in the cloud, the data are encrypted, and to download the data, the client should use the key to decrypt the data. This scheme provides better performance, whereas the speed of encryption is slow.

Yang et al. [16] proposed File Remotely Keyed Encryption and Data Protection (FREDP). Here, it involves the communication between the mobile terminal, the private cloud, and the public cloud. In this method, storage and the computation load of the mobile terminal are reduced. The huge amount of data and the computation task are forwarded to the private clouds. This technique involves four phases: file remotely keyed encryption, ciphertext uploading, storage phase, and data integrity verification. This method is secure, but file sharing is slow.

Li et al. [17] proposed Extended File Hierarchy Attribute-Based Encryption. Here, cloud environment saves the ciphertext in storage space and minimizes the computation encryption load. It encrypts multiple files in the same access itself. Extended File Hierarchy access tree is used to adopt the implementation of File Hierarchy Attribute-Based Encryption (FH-ABE). The steps involved in this are setup, encrypt, keygen, and decrypt. This method proves well in security, whereas the authority center in this scheme is not trustworthy.

Sharma and Kalra [18] proposed an authentication pattern based on a quantum distributed key for identity authentication in cloud computing. Here, the proposed authentication protocol consists of four phases. The first focuses on the registration phase; the server records the authentication parameter of the user. The second is the login phase, to use the services of the cloud server, the user sends a request to the server. The third is the authentication phase; the user and cloud server equally authenticate with other. The password is last phase; the user changes the password. An authentication scheme involves two entities: a user who wants cloud services and a service provider server. The server is a reliable authenticated authority, and the identity of the user must be verified when he communicates with or logins to a network. However, this method has high security, but it has computational complexity.

Wu et al. [19] proposed secure, searchable public-key encryption with private protection (SPE-PP). Here, it uses Diffie-Hellmen to generate the secret key. This system consists of four entries: Certificate authority, Data sender, Data user, and CSP. This method utilizes the following algorithms: setup, keygen algorithm, SPE-PP algorithm, and trapdoor algorithm. The performance of this scheme is better in terms of security. Moreover, the time taken for the encryption is quite high and the system is complex.

Sharma et al. [20] proposed a hybrid cryptographic approach for file storage mechanism in cloud computing. This proposed hybrid encryption method involves RC4, AES, and DES. The procedure in this scheme is as follows: a file is partitioned into three portions and sent to AES, DES, and RC4. The generated ciphertexts are merged, and ciphertexts are downloaded from cloud and decrypted. In the decryption reverse process of encryption takes place. It gives better execution times. However, security is quite low.

2.3 PROBLEM DEFINITION

From the literature review, it is concluded that in existing methods, the unauthorized user can easily edit the data in the cloud without the knowledge of the data owner and takes more computation time for the cryptography process. Furthermore, it is difficult to create a secure system in the cloud due to unauthorized users. For authentication of data transactions, OTP schemes and barcode are used. The memory space of barcode is not enough to store the data. If the user is in offline mode, it is quite difficult to send OTP, which slows down the process. If the data processing depends on the third-party organization, hackers gather the information about the user. Due to these issues, developing a new security scheme that provides both data confidentiality and data integrity is necessary.

A Hybrid Cryptography Technique with Blockchain

2.4 OBJECTIVES

The objectives of this research work are as follows:

- The main objective is to design a secure cloud storage system.
- To enhance the authentication in cloud environment, biometric authentication is used for each customer who wants to access the cloud environment so that only legitimate user can access the cloud services.
- To enhance the security and diminish the execution time using hybrid cryptographic algorithm.
- To provide secure data transaction and data integrity using data matrix code and blockchain verification scheme.
- To show the dominance of the proposed approach by comparing it with existing methods.

2.5 PROPOSED METHODOLOGY

Traditionally, the sensitive data are stored in a centralized environment which is untrustworthy. The data owner has to believe the third party, i.e., centralize the system to maintain the security and privacy of the data. If the hackers hack the third-party (centralized) system, then the security has to be compromised. In addition, there is a communication overhead that happens between the owner and the third-party system. In order to overcome these issues, the proposed approach does not support any third-party administrator (TPA), and so it offers scalability and flexibility.

The architecture of the proposed method is shown in Figure 2.1. The user who needs to utilize the cloud environment must enroll his username, password, and E-mail ID

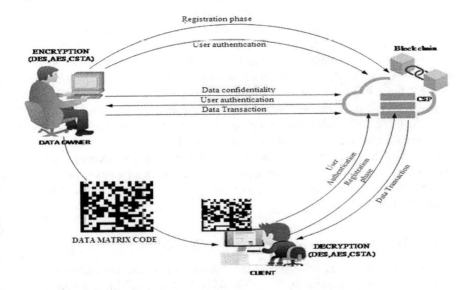

FIGURE 2.1 Architecture of the proposed method.

along with their iris. The authenticated client can upload the encrypted file in CSP. The client can retrieve the file with the knowledge of data owner. The data encryption file is selected and divided into three equal parts using the file system module. Then, each part is encrypted simultaneously by hybrid cryptographic algorithms such as DES, AES, and CST. This increases the security level and reduces the encryption time and decryption time over various file sizes compared to a single cryptography algorithm. Then, encrypted parts are merged and saved into a single file and is uploaded on the cloud servers. Moreover, to enhance the authentication, a data matrix code is used. The data stored by the user in the cloud should have data integrity and security, so we built a blockchain-based data verification technique that consists of setup and check proof stages. This provides high security against data loss caused by the hacker.

2.5.1 Registration Phase

Authentication plays a vital role for many tasks. User authentication is used to check the user's identity, i.e., verify the data of the correct sender. Simply, it is the process of confirming the verification of the user. In the cloud environment, any user can act as a legitimate person so that it may be a chance for unauthorized access occurring at any time. The biometric verification process is used to avoid this unauthorized access [21]. The biometric verification technique is extremely acceptable around the world and is easy to use. Businesses using the biometric verification technique in the cloud environment make the cloud system more secure, and only authorized users can access the cloud environment. In this proposed scheme, iris authentication is used to communicate between authenticated persons. Biometric authentication, such as iris, face, and fingerprint, is a more effective technique than password authentication scheme. During the registration phase, the iris scanner device captures the images of one or both of the human iris and stores the result in the CSP along with their user credentials. Iris recognition technology provides secure and reliable services as compared to all the biometric verification methods. It also includes an error-free verification process.

During the uploading phase, the iris scanner device captures the images of the data owner's iris and sends the details to the CSP. The CSP authenticates the data owner by comparing the iris of the data owner with the already stored iris image in the CSP. Once it is valid, the CSP allows the data owner to upload their file in the cloud environment. If the cloud user wants to retrieve the file, then he/she registers their iris using iris scanner device and sends to the CSP. Then the CSP compares the client's iris image with the already stored one. If matches occur, then the CSP informs the data owner that the requested client is legitimate and secure to exchange information with them. After the authentication is completed by biometric, the data owner decides whether communication with that client continues or not without using any third-party system.

2.5.2 Data Confidentiality Using Hybrid Algorithm

Data stored in the cloud environment must be free from unauthorized access, i.e., cloud users must ensure that their data should be kept secure. This is accomplished by using the cryptographic method. Data confidentiality is used to protect the data against unauthorized access or disclosure. Cryptography is a process of securing information by generating secret code using mathematical concepts and procedures. It is a technique that is used to perform encryption and decryption of data. Cryptography converts the plaintext into a scrambled message wherein normal users cannot read the message. There are two different types of cryptographic techniques: symmetric cryptography and asymmetric cryptography. Symmetric cryptography performs encryption and decryption using the shared key, whereas asymmetric cryptography uses one key for encryption and another key for decryption. The symmetric encryption technique is preferred to use when a large amount of data is required to exchange. In this proposed system, the hybrid algorithm is used to encrypt and decrypt the content.

A hybrid cryptographic algorithm is an efficient algorithm that combines the desired features of each algorithm so that the algorithm's overall performance is better than individual components. The hybrid encryption algorithm is a symmetric block cipher consisting of three different symmetric algorithms such as AES, DES, and CST. AES [22] consists of a symmetric block cipher where a shared key is used for both encryption and decryption. The input plaintext consists of 128 bits and key size consists of 128, 192, or 256 bits. AES generates key for each round using the key expansion method. The input plaintext is processed as state matrix which is passed through rounds and finally produces a ciphertext. Each round follows various steps such as substitute bytes, shift rows, mix columns, and Add round key. DES [23] is a symmetric block cipher which consists of 64 bits of input plaintext and 64 bits of key value. However, only 56 bits are used by DES algorithm and the remaining 8 bit is used as a parity bit and therefore discarded. CST [24] consists of symmetric block cipher, which accomplishes the partition and then arranges them into matrix format and performs shifting operations such as column shift, row shift, primary diagonal shift, and secondary diagonal shift and finally produces the ciphertext.

The input plaintext is divided into "n" number of blocks b_i in the hybrid cryptographic process. Then it is divided into three parts. The first part m_1 consists of $\left(0 : n/3 - 1\right)$ blocks, the second part m_2 consists of $\left(n/3 : 2n/3 - 1\right)$ blocks, and third part m_3 consists of $\left(2n/3 : n - 1\right)$ blocks. The first part is encrypted using AES encryption technique. The second part is encrypted using DES, and the last part is encrypted using CST. All three parts are executed in parallel. Finally, the ciphertext, which combines three resultant ciphertexts, is sent to cloud storage.

Algorithm 1: Hybrid Algorithm for Encryption

Initialization

Step 1: Partition the content of the file into three parts
Step 2: Apply AES [22] the encryption scheme on the first part of the content
    Step 2.1: The first part of the message can be defined as

$$m_i = \sum_{i=0}^{n/3-1} b_i, 0 \leq i \leq n/3 - 1$$

Step 2.2: for $i = 0$ to $n/3 - 1$

$$C_i = E_{AES}(K, m_i)$$

Step 3: Apply DES [25] encryption scheme on the second part of the content
Step 3.1: The second part of the message can be defined as

$$m_i = \sum_{i=n/3}^{2n/3-1} b_i \frac{n}{3} \leq i \leq 2n/3 - 1$$

Step 3.2: for $i = \frac{n}{3}$ to $2n/3 - 1$

$$C_i = E_{DES}(K, m_i)$$

Step 4: Apply CST [24] encryption scheme on the third part of the content
Step 4.1: Third part of the message can be defined as

$$m_i = \sum_{i=2n/3}^{n-1} b_i \frac{2n}{3} \leq i \leq n - 1$$

Step 4.2: for $i = \frac{2n}{3}$ to $n - 1$

$$C_i = E_{CST}(K, m_i)$$

Step 5: Combine three parts of encrypted content

$$C = \sum_{i=0}^{n-1} c_i$$

In the decryption process, the ciphertext is separated into three parts. The AES decryption technique has been applied on the first part, the DES decryption process on the second, and CST decryption process on the last part, then combine three parts of the final messages altogether, which is the original input as entered by the data owner.

Algorithm 2: Hybrid Algorithm for Decryption

Initialization

Step 1: Partition the content of the file into three parts
Step 2: Apply AES [22] decryption scheme on the first part of the content

A Hybrid Cryptography Technique with Blockchain

Step 2.1: The first part of the message can be defined as

$$m_i = \sum_{i=0}^{\frac{n}{3}-1} b_i, 0 \leq i \leq n/3 - 1$$

Step 2.2: for $i=0$ to $n/3 - 1$

$$m_i = D_{AES}(K, C_i)$$

Step 3: Apply DES [25] decryption scheme on the second part of the content

Step 3.1: The second part of the message can be defined as

$$m_i = \sum_{i=n/3}^{2n/3-1} b_i, n/3 \leq i \leq 2n/3 - 1$$

Step 3.2: for $i = \dfrac{n}{3}$ to $2n/3 - 1$

$$m_i = D_{DES}(K, C_i)$$

Step 4: Apply CST [24] decryption scheme on the third part of the content

Step 4.1: Third part of the message can be defined as

$$m_i = \sum_{i=2n/3}^{n-1} b_i, \dfrac{2n}{3} \leq i \leq n - 1$$

Step 4.2: for $i = \dfrac{2n}{3}$ to $n-1$

$$m_i = D_{CST}(K, C_i)$$

Step 5: Combine three parts of the decrypted content

$$M = \sum_{i=0}^{n-1} m_i$$

The proposed hybrid algorithm makes it difficult for an intruder to decrypt the ciphertext. This algorithm provides fast processing techniques because each block of the content encrypts parallelly and provides high security. This scheme decreases processing overhead and achieves lower energy consumption [26,27].

2.5.3 SECURE DATA INTEGRITY AND THE TRANSACTION

Data integrity provides assurance, accuracy, and consistency for the data. It refers to the trustworthiness of data. It is necessary to maintain data integrity in the cloud environment. Otherwise, there will be a loss of over millions of dollars in businesses [28,29]. The blockchain is an important solution for data integrity. Blockchain ledger cannot be edited and added by an unauthorized user because it involves timestamp so that the history of data is auditing frequently. Blockchain consists of a distributed database of records where each block of data is chained together through chronological order [30,31]. The blockchain has unique hash value corresponding to the file generated by the Merkle hash tree. The proposed scheme is used to check the integrity using a secured ledger without support on a TPA. The owner and CSP maintain the ledger in a secured manner for attesting the correctness of outsourced data. In this scheme, data owner stores the encrypted data and Hashtag of the complete data on the blockchain distributed database in the CSP. For verification purposes, data owner also keeps the Hashtag of the data. The blockchain-based data verification technique consists of two phases: the setup and check proof phases.

2.5.3.1 The Setup Phase

In this phase, the data owner generates a Hashtag for each block of data D. Each block header consists of four parts: (i) hash of the current block $H_i(D)$ where $i = 1$, (ii) timestamp (TS), (iii) the Hashtag of the previous block $H_i-1(D)$ where $i = 2,3...n$, (iv) random integer (nonce); data owner uploads the Hashtag of the data to cloud storage. After receiving the data, CSP runs the same hash function for the data and compares the resultant hash function with the received hash function. If it matches, the CSP sends the confirmation to the data owner. Otherwise, it sends an error message, which means that data may get lost at the time of data transfer. At last, CSP and data owner publish the H(D) in the ledger.

Algorithm 3: Setup Phase

```
Input: Data, D
Calculates Hashtag
       H_i (D) = H_i ((D) || (TS) || N_i), i = 1
                H_{i-1} ((D) || (TS) || N_i), i = 2,3...n
Upload the data to CSP
CSP computes the same Hashtag for the data such as H' (D)
For i = 1 to n
    If (H'_i (D) == H_i (D))
       Return 'confirmation message' to data owner
       Adds Hashtag in the ledger of CSP and Owner
    else
       return 'error during data transfer'
End for
End
```

2.5.3.2 Check Proof Phase

In this phase, the client checks the proof of the file and retrieves the file with the knowledge of the data owner. The CSP authenticates the client using the iris recognition mechanism and intimates that the client is legitimate. If the data owner wishes to communicate with the client, then he/she instructs the CSP to share the Hashtag details to the client. Data owner sends data matrix code to the client wherein the data matrix hides the copy of Hashtag. The client verifies whether the copy of the Hashtag sent by the CSP matches with the copy sent by the owner using data matrix code. If it is valid, then it represents that the unauthorized user did not access the file and hence it is protected.

Algorithm 4: Check Proof Phase

Input: The Hashtag of the current block, Hashtag for the previous block, timestamp and nonce
Output: true (no data tampered), false(data tampered)
If ((authenticates (client)== yes))
Data owner intimate the CSP to send the hash function details to the client
End
CSP send the hash function for the data to the client
Data owner also send the hash function using data matrix code to the client
Client verifies
If ($H_{CSP}(D) == H_{OWNER}(D)$)
Return true
Else
Return false
End

Any user can access the data in the public cloud environment at any time; there is a chance for hackers to enter the CSP [32,33]. Nowadays, encryption is not a solution to protect the data from unauthorized attacks; it is also essential to protect the data from corruption. Blockchain technology is used to acquire data integrity and make sure the security of the data owner. It is mainly used to identify whether the data have been tampered or not. In this proposed system, the data owner involves in the auditing process so that he/she is conscious of the data in the cloud environment [34–36]. Here, the data owner will decide to whom they want to share the information, even though third-party CSP is available. Also, the data owner generates the data matrix code without using TPA. Hence data ownership is provided in this scheme. An unauthorized user cannot view information stored behind the data matrix code, so it is secure.

2.6 PERFORMANCE ANALYSIS

In this section, the proposed hybrid cryptography method for data security tested in Simulink and its performance is compared with the existing method in terms of encryption time and decryption time to demonstrate its effectiveness.

2.6.1 THE SIMULATION RESULTS

The data owner sends the encrypted copy of the file and signature to the cloud environment. The ciphertext and signature is stored into the CSP. The average time of the hybrid encryption and decryption algorithm is compared with the existing algorithm such as DES, AES, and CST for the different size of the data. This is represented in Figures 2.2 and 2.3 such as average encryption time for the hybrid cryptographic scheme and average decryption time for the hybrid cryptographic scheme.

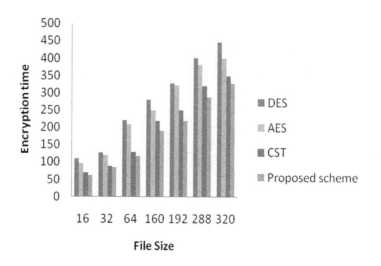

FIGURE 2.2 Average encryption time for hybrid cryptographic scheme.

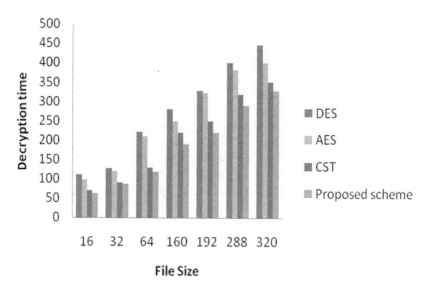

FIGURE 2.3 Average decryption time for hybrid cryptographic scheme.

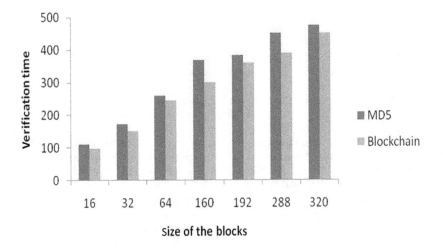

FIGURE 2.4 Signature verification time for the proposed scheme.

2.6.2 SIGNATURE VERIFICATION

Data owner generates and verifies the data's signature by using blockchain technique and sends into the cloud environment. In this scheme, the blockchain signature verification is compared with the existing signature verification algorithm such as Message-Digest algorithm 5 (MD5) for the varying size of the block. This is represented in Figure 2.4, such as signature verification time for the proposed scheme.

2.7 CONCLUSION

In this scheme, highly secured blockchain techniques are used for auditing the data integrity without using any third-party auditor. Hybrid cryptographic algorithm secures the data in the cloud environment. The data cannot be retrieved by the client from the CSP without the knowledge of the data owner. Data transaction can be secured by using data matrix code. In order to make the cloud environment more secure and safe, iris authentication method is used between sender and receiver to authenticate the authorized user. The performance analysis shows that the proposed scheme is secure and takes less communication time than the existing one.

REFERENCES

1. Mell, P, and Grance T. 2011, "The NIST definition of cloud computing", (Draft). NIST. http://www.production scale.com/ home/2011/8/7/the-nist-definition-of-cloud-computingdraft.html#axz z1X0x KZRuf.
2. Apostu A, Puican F, Ularu G, Suciu G, and Todoran G. 2013, "Study on advantages and disadvantages of cloud computing – the advantages of telemetry applications in the cloud", *WSEAS Conference Proceedings in Applied Computer Science and Digital Services*, (pp. 118–123).

3. Carroll M, Van Der Merwe A, and Kotze, P. 2011, "Secure cloud computing: benefits, risks and controls", *IEEE Conference on Information Security South Africa*, (pp. 1–9), ISBN: 978-1-4577-1483-2.
4. Xue CTS, and Xin FTW. 2016, "Benefits and challenges of the adoption of cloud computing in business", *International Journal on Cloud Computing: Services and Architecture*, vol. 6, no. 6, pp. 1–5.
5. Alsmadi D, and Prybutok V. 2018, "Sharing and storage behavior via cloud computing: Security and privacy in research and practice", *Computers in Human Behavior*, vol. 85, pp: 218–226.
6. Cheng L, Van Dongen B, and Van Der Aalst W. 2019, "Scalable discovery of hybrid process models in a cloud computing environment", *IEEE Transactions on Services Computing*, vol. 13, no. 2, pp. 368–380. doi: 10.1109/TSC.2019.2906203
7. Amin R, Kumar N, Biswas GP, Iqbal R, and Chang V. 2018, "A light weight authentication protocol for IoT-enabled devices in distributed cloud computing environment", *Future Generation Computer Systems*, vol. 78, pp: 1005–1009.
8. Zhou L, Li X, Yeh KH, Su C, and Chiu W. 2019, "Lightweight IoT-based authentication scheme in cloud computing circumstance", *Future Generation Computer Systems* vol. 91, pp. 244–251.
9. Namasudra S, Devi D, Kadry S, Sundarasekar R, and Shanthini A. 2020, "Towards DNA based data security in the cloud computing environment", *Computer Communications*, vol. 151, 539–547.
10. Wang Y, Miao M, Shen J, and Wang J. 2019, "Towards efficient privacy-preserving encrypted image search in cloud computing", *Soft Computing*, vol. 29 no. 6, pp. 2101–2112.
11. El Makkaoui K, Beni-Hssane A, and Ezzati A. 2019, "Speedy Cloud-RSA homomorphic scheme for preserving data confidentiality in cloud computing", *Journal of Ambient Intelligence and Humanized Computing*, pp: 4629–4640.
12. Akande NO, Abikoye CO, Adebiyi MO, Kayode AA, Adegun AA, and Ogundokun RO. 2019 "Electronic medical information encryption using modified blowfish algorithm", *International Conference on Computational Science and Its Applications*, (pp. 166–179). Springer, Cham.
13. Prakash R, Chithaluru P, Sharma D, and Srikanth P. 2019, "Implementation of trapdoor functionality to two-layer encryption and decryption by using RSA-AES cryptography algorithms", in Nath V, and Mandal J (eds.), *Nano electronics, Circuits and Communication Systems*, (pp. 89–95). Springer, Singapore.
14. Goyal V, and Kant C. 2018, "An effective hybrid encryption algorithm for ensuring cloud data security", in Aggarwal V, Bhatnagar V, and Mishra D (eds.), *Big Data Analytics*, (pp. 195–210). Springer, Singapore.
15. Lee BH, Dewi EK, and Wajdi MF. 2018, "Data security in cloud computing using AES under HEROKU cloud", *IEEE Wireless and Optical Communication Conference (WOCC)*, (pp. 1–5).
16. Yang Y, Chen X, Chen H, and Du X. 2018, "Improving privacy and security in decentralizing multi-authority attribute-based encryption in cloud computing", *IEEE Access*, vol. 6, pp. 18009–18021.
17. Li J, Chen N, and Zhang Y. 2019. "Extended file hierarchy access control scheme with attribute based encryption in cloud computing", *IEEE Transactions on Emerging Topics in Computing*, vol. 9, no. 2, pp. 983–993.
18. Sharma G, and Kalra S. 2018, "Identity based secure authentication scheme based on quantum key distribution for cloud computing", *Peer-to-Peer Networking and Applications*, vol. 11, pp: 220–34.
19. Wu L, Chen B, Zeadally S, and He D. 2018, " An efficient and secure searchable public key encryption scheme with privacy protection for cloud storage", *Soft Computing*, vol. 22 no. 23, pp. 7685–96.

20. Sharma S, Singla K, Rathee G, and Saini H. 2020, "A hybrid cryptographic technique for file storage mechanism over cloud", in Luhach A, Kosa J, Poonia R, Gao XZ, and Singh D (eds.), *First International Conference on Sustainable Technologies for Computational Intelligence* (pp. 241–256). Springer, Singapore.
21. Naveed G, and Batool R. 2015, "Biometric authentication in cloud computing", *Journal of Biometrics & Biostatistics*, vol. 6, no. 5, pp: 1000258–1000258, ISSN: 2155-6180.
22. Saha R, Geetha G, Kumar G, and Kim T-h. 2018, "RK-AES: An improved version of AES using a new key generation process with random keys", *Security and Communication Networks*, vol. 2018, pp: 1–11.
23. Patel, JR, Bansode RS, and Kaul V. 2012, "Hybrid security algorithms for data transmission using AES-DES", *International Journal of Applied Information Systems (IJAIS)*, vol. 2, no. 2, pp: 16–21.
24. Selvi K, and Kavitha V. 2012, "Crypto system based authentication using CSTA in grid", *International Journal of Computer Application*, vol. 48, no. 22, pp: 45–51, ISSN 0975-8887.
25. Moharir M, and Suresh AV. 2014, "Data security with hybrid Aes-Des", *Elixir Computer Science & Engineering*, Vol. 66, pp. 20924–20926.
26. Rizk R, and Alkady, Y. 2015, "Two-phase hybrid cryptography algorithm for wireless sensor networks", *Journal of Electrical Systems and Information Technology*, vol. 2, pp: 296–313.
27. Susarla S, and Borkar G. 2014, "Hybrid encryption system", *International Journal of Computer Science and Information Technologies*, vol. 5, no. 6, pp. 7563–7566.
28. Yadav AK, Bharti R, and Raw RS. 2021, "SA^2-MCD: Secured architecture for allocation of virtual machine in multitenant cloud databases," *Big Data Research: An International Journal*, vol. 24. Doi: 10.1016/j.bdr.2021.100187.
29. Yadav AK, Bharti R, and Raw RS. 2018, "Security solution to prevent data leakage over multitenant cloud infrastructure", *International Journal of Pure and Applied Mathematics*, vol. 118 no. 07, pp 269–276.
30. Teshome A, Peisert S, Rilling L, and Morin C. 2019 "Blockchain as a trusted component in cloud SLA verification", *Proceedings of the 12th IEEE/ACM International Conference on Utility and Cloud Computing Companion*, (pp. 93–100), New York.
31. Li S, Liu J, Yang G, and Han J. 2020, "A blockchain-based public auditing scheme for cloud storage environment without trusted auditors", *Wireless Communications and Mobile Computing*, vol. 2020, pp: 1–13.
32. Kumar M, Yadav AK, and Raw RS. 2018, "Global host allocation policy for virtual machine in cloud computing", *International Journal of Information Technology*, vol. 10, pp. 279–287.
33. Ahmed A, Abdul Hanan A, Omprakash K, Lobiyal DK, and Raw RS. 2017, "Cloud computing in VANETs: Architecture, taxonomy and challenges", *IETE Technical Review*, vol. 35 no. 5, pp. 523–547.
34. Kamal R, Saxena NG, and Raw RS. 2016, "Implementation of security & challenges on vehicular cloud networks", *International Conference on Communication and Computing Systems (ICCCS-2016)*, (pp. 379–383), Doi: 10.1201/9781315364094-68.
35. Dixit A, Yadav AK and Raw RS. 2016, "A comparative analysis of load balancing techniques in cloud computing", *International Conference on Communication and Computing Systems (ICCCS-2016)*, (pp. 373–377), Doi: 10.1201/9781315364094-67.
36. Parween D, Yadav AK and Raw RS. 2016, "Secure architecture for data leakage detection and prevention in fog computing environment", *International Conference on Communication and Computing Systems (ICCCS-2016)*, (pp. 763–768), Doi: 10.1201/9781315364094-137.

3 Fog Computing Environment in Flying Ad-hoc Networks
Concept, Framework, Challenges, and Applications

Devraj
GGS Indraprastha University

Ram Shringar Rao
Netaji Subhas University of Technology

Sanjoy Das
Indra Gandhi National Tribal University-
Regional Campus Manipur

CONTENTS

3.1 Introduction 31
      3.1.1 Motivation 33
      3.1.2 Organization 33
3.2 Fog Computing 34
3.3 UAV-Based Fog Computing 35
3.4 Framework and Architecture of UAV-Based Fog 37
3.5 Challenges for UAV-Based Fog 40
3.6 Applications and Scope of UAV-Based Fogs 42
3.7 Techniques for Implementation and Experiments 44
3.8 Conclusion 45
References 45

3.1 INTRODUCTION

The mobile ad-hoc network comprises mobile hosts equipped with remote communication devices. The message is relayed from a host to another host in this network; the hosts behave like routers [1]. The transmission of the mobile host can be received by all hosts which are inside its transmission range because the broadcast had the nature of the wireless correspondence or omni-directional reception antennae. But suppose there are two remote hosts which are out of their transmission range in the ad-hoc network. In that case, the other mobile hosts which are situated between them can advance messages, which adequately build the associated networks between the portable hosts in the conveyed territory.

Vehicular Ad-hoc Networks (VANETs) is a type of network made from the idea of building up a network of vehicles for specific need or circumstance. VANET can provide security and other types of applications to passengers [2]. VANET has now been set as a dependable network that the vehicles can be used for communication on highways or metropolitan scenarios. The VANETs provide the basic operational framework which is required by the modern Intelligent Transportation Systems (ITS) [3,4]. VANET has now been set as a reliable network that vehicles can use for communication purposes on the road. Alongside the benefits, there can be an enormous number of challenges in VANET, like arrangement of high network, Quality of Services (QoS) and data transmission, and the privacy and security to vehicles. The cloud of vehicles can also be formed, known as the Vehicular Cloud Network (VCN), which sends the messages to the vehicles on time [5].

Flying Ad-hoc Network (FANET) is one of the new fields of the Wireless Ad-hoc Network, which can establish a connection between portable and small devices like Unmanned Aerial Vehicles (UAVs) [6]. As of now, UAVs have been utilized in different applications, for example, observing [7], surveillance [8], and geography [9]. This clarifies the UAVs' capacity to perform the complex operations and move adaptability and ease of flight. In ordinary missions, the administrator can control the position of the vehicle in each circumstance. For some cases, when there is a semiautonomous mission, the administrator is answerable for only some assignments, such as landing and taking off the airplane. In that sense, the aerial robots are associated with a supervised system (ground stations—GS) that are normally situated at the cloud and is liable for all high-level handling. Though, this cloud-based methodology can be improper for the sensitive real-time machines once the exchange information among the devices would produce the absence of mobility, the greater cost of communication bandwidth, energy requirements of embedded systems, communication delay, and data excess [10].

In completely independent reactive missions, the direction can be installed onboard and the mission can be performed without the administrator nearby. There is another pattern for the computing paradigm to moderate these issues to make the computing and capacity nearer to end-devices, which in this case are the robots [11]. The fog computing emerges as a middle-of-the-road layer among end-devices and cloud to improve power utilization, latency, efficiency, and adaptability. This innovation permits defeating the limitation of the centralized cloud computation with empowering information obtaining, handling, and capacity on fog devices [12]. By these presumptions, this chapter is proposing a fog-based system centered with respect to UAVs topology. This methodology utilizes a UAV as a fog processing hub to offer types of services. These services should be sent by this node alongside clustering and filtering. The filtering technique has a significance-based classifier that permits basic data to be conveyed as per the application requirement. The fog processing can likewise be a coordination and supervision system that isn't resented in different architecture.

3.1.1 Motivation

Network arrangement between the UAVs is very irregular because of UAVs' high mobility degree [13,14]. There are a few applications which are very subject to delays. For instance, Search and Rescue and the inspection are commonly executed at remote areas with low correspondence assets. In this specific situation, greater part of the choices is team coordination [15] and target detection [16]. These are the tasks which are especially delay-sensitive like, for example, in the target identification, the item can be misplaced in portions of seconds if there should arise an occurrence of the improper detection. Other than these restrictions, many applications require lots of information, particularly for streaming the videos to the GS for monitoring or image processing [17]. The research introduced in [18] recommends that the one second of delay is a test in the cloud-supported applications and lower than 100 ms are unreachable. These issues impact real-time applications and diminish the ability to control systems. A few applications with the UAVs are subject to this sort of issue, which can degrade the nature of the mission exposed to the delays. The places with low communication structures can bring data transmission challenges that must be appropriately addressed. There is another significant motivation in the fog and cloud power consumption. The amount of accessible energy fundamentally restricts the airplane's mission services and time; quad-rotors regularly have flight times lower than 25 minutes. In that sense, any advancement can improve the system in general execution. So the inspiration for the current work's improvement comes from the non-usual fog distributed computing appropriateness with different UAVs. Now the principal challenge is to propose an architecture that is used to assess the relevance of fog distributed computing collaboration targeting at optimizing latency while maintaining power utilization and throughput under a reach. Based on discussed issues, this work can address the following research challenges:

1. A stage to help the fog distributed computing convey implied least throughput and power utilization.
2. A representation to investigate the productivity of the fog cloud computing coordinated effort and also its prerequisites for UAVs.
3. Data handling nearer to end-devices.

3.1.2 Organization

The rest of the chapter is coordinated as follows. The related works and foundation to fog computing are introduced in Section 3.2. Section 3.3 examines focal points that can be upgraded when applying fog processing to the UAV-based FANET. The architecture of UAV-based fog is shown in Section 3.4. In Section 3.5 we introduce different types of challenges which can occur while applying fog computing in FANET. The scope of UAV-based fog computing is presented in Section 3.6. In Section 3.7, we introduce few techniques which different researchers used to implement UAV-based fog computing. The conclusion and closing comments are in Section 3.8.

3.2 FOG COMPUTING

The majority of the IoT applications should connect with different networks or cloud for various purposes. The IoT applications may use any cloud to store information, dominant handling, or to connect with advanced services that can execute an intelligent system to schedule, optimize, discover knowledge, or plan. Cloud computing gained a lot of interest in vehicular communication due to the architectural similarity between mobile cloud computing and VANETs [19,20]. A model uses services and features given by the cloud, for example, elastic sources at very minor cost. Nonetheless, interfacing IoT application to cloud has numerous limitations because the cloud can't control a portion of basic characteristics and the requirements of the IoT applications, like low latency responses, exceptionally heterogeneous devices, location awareness, and portability. To overcome these drawbacks, Cisco presented the idea of fog processing [21]. The structural design of coordinating fog processing, IoT has appeared in Figure 3.1. The architecture shows that the fog gave more restricted control, real-time monitoring, and an enhancement to IoT applications, and cloud can give global control, improvement, flexible recourses, planning, and extra superior services used by the IoT applications.

The research work is done to overview, assemble, study, and investigate fog processing. Yi et al. [22] and Stojmenovic and Wen [23] overviewed fog processing advantages and issues. Hong et al. [24] designed a programming model for different fog computing stages. Satyanarayanan et al. [25] planned Cloud lets, a stage that has comparable capacities to fog processing. Dasgupta et al. [26] proposed Para-Drop, a platform for fog that is accessible over any wirelessly connected router. Zhu et al. [27] projected the utilization of a fog websites execution improvement. Ha et al. [28] utilized the Cloud lets to plan and implement the real-time wearable, which can be cognitive help on the Google Glass. Cao et al. [29] utilized fog computing to empower the analytics to pervasive applications like health monitoring. Hassan et al. [30] researched utilizing the fog computing to handle off-load and capacity development for the mobile application. The issues like protection and security in fog computing has been concentrated by Dsouza et al. [31] and Stojmenovic et al. [32].

Fog computing can help to improve the cloud computing worldview by giving the small platforms situated on the edges of the networks which is nearer to the networks and IoT devices. These are the platforms that can work the cloud-like service to help the IoT activities. These types of services can be used in communication, control, handling, setup, observation, estimation, and supervision services to help specific IoT applications. By utilizing fog computing, any application in a particular zone can use the architecture that utilizes a committed Personal Computer (PC) accessible locally or at least customers or the edge devices close by. The fog stage encourages the executing services geologically close by IoT applications and simultaneously can be utilized by cloud services. This gives various advantages to IoT applications like [33]:

- To offer area alert services.
- To provide low latency services.
- To support the services like streaming processing and correspondence.
- To provide the improved hold for the widely distributed applications.
- To provide more productive correspondence in different systems.
- To offer better QoS (Quality of Services) control.
- To support the improved access control.

3.3 UAV-BASED FOG COMPUTING

This segment includes the UAV-based fog, i.e., fog computing in FANETs. Unlike a couple of years back, the only motivating factor was the military applications following the advancement of the UAV-based system; many civil applications have shown up and contributed to driving the improvement of the UAV innovations' advance [34]. UAVs can play significant parts in numerous applications, such as SAR, monitoring environment-related events, and awareness of natural disasters. The UAVs can provide lots of applications compared to normal manned aerial vehicles. Small and medium cheap automated airplanes are presently commercially accessible. Some of these convey information through cameras, sensors, actuators, and storage and specialized devices. The UAVs can easily reach locations and provide information that is very hard for access. UAVs can also decrease operational expenses while there is an improvement in efficiency [35].

There are a lot of repetitive and important tasks which could be automatic. Also, using the normal manned aerial vehicle in a dangerous mission can expose the pilot to a very high risk, but this risk can be reduced by using the UAV. UAV-based fog is a type of UAV that can work as the fog node. If there is a need for any of the fog nodes to help any IoT applications, UAV-based fog can be allocated. In addition to that, it could also work like any gateway which can handle the integration between cloud and IoT devices. Each UAV-based fog can have storage devices, processors, a fog computational stage, and communication devices. The fog stage can also execute a lot of operations to help the different IoT applications. The UAV-based fog is shown in Figure 3.1; here, it is located at the area of IoT. It enables the integration between the cloud and IoT applications. UAV-based fog in that particular region also provides a set of low latency operation for IoT applications.

UAV-based fog conveys close to IoT device, the communication which conveys information between UAV-based fog and IoT devices could be handled by Personal Area Network classes like IEEE 802.15.4 (Zigbee) or 801.15.1 (Bluetooth).

FIGURE 3.1 An IoT application supported by UAV-based fog.

These protocols are generally used for lower bandwidth, low energy consumption, and short-range. On the other hand, it can be handled by the protocols that operate on longer range, for example, classes like IEEE 802.11 (Wi-Fi) or protocols operate in the LAN. In distinction, the communication among UAV-based fog and cloud or some other system has to utilize some protocol that belongs to the Wireless Area Network (WAN) class, such as Worldwide Interoperability for Microwave Access (WiMAX) (IEEE 802.16), satellite communication, or cellular. The protocol selection can be done on the basis of its location and the IoT applications which are being supported. For instance, both cellular and WiMAX can be used within cities and regions, and it cannot be used in remote locations because of no cellular coverage or the WiMAX is inaccessible. So, satellite communication could be the better option to establish the connection between the cloud and the UAV-based fogs in these conditions. A lot of opportunities can be provided by UAV-based fogs to some of the IoT applications. Some of these opportunities are listed:

- Flexibility: Multiple UAV-based fogs can have capabilities that are heterogeneous, as exposed in Figure 3.2. These types of heterogeneous UAV-based fogs can be planned to accomplish IoT application requirements. For instance, if a high processing capability is required near the IoT area but the requirement of communication is limited with the cloud, then one or more UAV-based fogs with powerful processors can be used there. In addition to that, more UAV-based fogs can be sent to support if there is a temporary requirement for more processing power or storage.
- Less Cost: As UAV-based fogs can be heterogeneous and rapidly deployed, cost-effective solutions for IoT applications can be provided.
- Fast Deployment: The deployment of UAV-based fogs can be easily done at any region to support the IoT applications. They can be deployed in those regions which are unreachable for manned vehicles or humans. The UAV-based fog can be easily deployed for the replacement of a faulty fog for the continuation of supporting operations in the IoT application.
- Scalability: A lot of UAV-based fogs can be deployed to support large-scale IoT applications. These fogs can be owed to assure few specific requirements for IoT applications. More UAV-based fogs can easily be sent to help as the size of IoT applications increases.

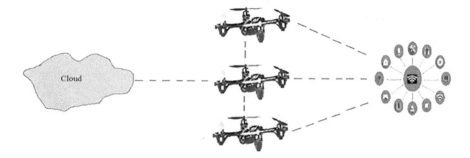

FIGURE 3.2 An IoT application supported by multiple UAV-based fogs.

- One UAV-based fog has the capability of communication with cloud, while the rest supports other features.

3.4 FRAMEWORK AND ARCHITECTURE OF UAV-BASED FOG

The programming models and diverse proposed stages like those which are already discussed can be customized and used for UAV-based fog. In any case, this segment includes a service-oriented stage for UAV-based fog. The fundamental purpose is to present any virtual environment that should utilize for sending the IoT application after creation. This segment includes the functions that can be seen as a bunch of the services that could be utilized for the various IoT applications to control and build the execution. The services are grouped to environmental service and core service. The core service is developed explicitly for core activities of UAV-based fog, such as the agent, security, area aware service, and service invocation [34]. These services generally offer control for the entire system. The environmental service gives access to service given by the cloud; service given by various sensors, actuators, vehicles, UAV, robots, etc.; and service given by different distributed fogs. The cloud service can be Platform as a Service, Infrastructure as a Service, and Software as a Service, which can describe various services for the IoT applications, which include optimize and services used for simulation and data mining.

UAV-based fog administrations can communicate, configure, monitor, stream, control, store, estimate, manage, and prepare service [36]. The interface is given by the IoT device services to use device functionalities. Either a direct interface to get the original service given by cloud or an IoT device or a UAV-based fog can be given by the environmental services or present some of the additional value for the original services, such as adding security highlights and reliability. While on-board code can be executed by some IoT devices to offer a few types of services, controlling other types of IoT devices can be done basically by the UAV-based fog or microcontroller, which has some types of services that give interfaces for the functionality of the devices. UAV-based fog can utilize the proposed service, applications accessible on the cloud, or the IoT devices, for example, any robot requesting specific help from a specific application that is accessible on the cloud.

Major functions of the UAV-based fog stage are to empower the smooth operation between remote IoT devices service, local IoT service, local UAV-based fog service, cloud services, and remote UAV-based fog service to control IoT applications successfully. It ought to permit any application or service accessible on one of the UAV-based fog or on the cloud to use the services that are accessible in entire environment, which appears in Figure 3.3. Each of the layers expresses its interface which makes it accessible for the different types of services. Utilizing architecture for the UAV-based fog platform gave a flexible way by which it can deal with link accessible services. A loose coupling accomplishes the UAV-based fog platform among connecting services. UAV-based fog platform manages communication, invocations, services advertisement, and discovery. Moreover, it may also be utilized for the execution of collaborative services over other types of service-oriented processing type system and with different IoT applications.

Many of the core services can be integrated by the UAV-based fog platform; however, the primary services are invocation services, security services, and the area-based help. The following are the five types of basic services to guarantee successful use of the other accessible services and encourage the main function they offer.

A. Discovery of Integration Services and IoT Resources

UAV-based fogs can migrate or move to various areas to help IoT devices and IoT applications be available at assigned territory. It is important for this situation that the UAV-based fog can find the available IoT resources, areas, and abilities in the new territory. Various methodologies can be utilized to improve the setup and discovery. The best way to deal with UAV-based fogs is Universal Plug and Play method [37] for integration and discovery. Using that methodology configuration and combination of the all devices which are accessible can be effectively automated. Using IoT devices, the devices' GPS abilities or estimated area estimation strategies can be utilized for IoT [38].

B. *Services* provided by *Broker*

Every UAV-based fog has many broker services and a broker that are answerable to IoT devices' services enrollment, advertisement, UAV-based fog, and search. The local IoT devices will have corresponding services which can be enabled by using them. The enrollment of these services will

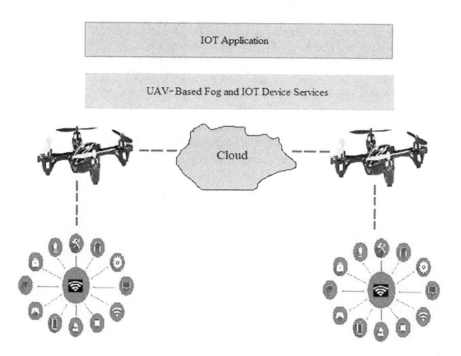

FIGURE 3.3 The service supplied by the different devices and platforms utilized by IoT applications.

be done with the broker. Each UAV-based fog has a local broker. The work of this local broker in UAV-based fog is just to keep up the data which is currently accessible services inside the UAV-based fog. This methodology is utilized to permit applications and services inside a UAV-based fog to use accessible services and resources and give low latency responses. Thus, the time expected to find services is minimized to efficiently and precisely use the services. The UAV-based fog broker will register the data in the environment about other accessible brokers. Those accessible brokers can be the brokers who have a place in the other UAV-based fogs or belong to the clouds. Any type of service or application that needs to utilize any specific service will look up service in local brokers. Other UAV-based fogs will keep up data about their services and local resources, including IoT devices currently connected with them.

In contrast, the cloud broker will keep up the data about service given by the cloud. The local broker can forward queries to known brokers if the local broker doesn't have the service. The UAV-based fog can also forward request sometimes to the cloud broker by which it can get service on the cloud. Every broker keeps up the data description utilizing the Web Service Description Language (WSDL) for the characterized service.

C. Remote or Local Services

The consumers start the services, remote and local with the UAV-based fog control. The remote service is the one that can be between a UAV-based fog service and another UAV-based fog service, between a UAV-based fog service and a cloud service, or between a UAV-based fog service and an IoT device service. UAV-based fog platforms address message addressing, message information marshaling or de-marshaling, and conveying request, responses, and communication connection between consumers-producers and executing services.

D. Area-Based Service

The UAV-based fog platforms can give location-based services. Contrasting to a normal service broker who is utilized over the Internet, the service broker for UAV-based fogs keep up extra data about the current places of at present associated IoT mobile devices. The main purpose behind keeping up current positions is that a portion of their services can be viewed as valuable and might be used just if the provider device is in a particular area; in any case, there is no use in using that service. One model utilizes a sensor service on a robot if the robot is accessible inside a particular area. A consumer in a UAV-based fog can look into a specific help inside a particular area through its UAV-based fog broker. If this service is accessible inside the UAV-based fog's reach, at that point, WSDL data about the service is sent to user to use the service. Otherwise, the service specialist of UAV-based fog can forward lookup request to the different brokers who are accessible on the cloud or other UAV-based fogs also if the service is accessible, at that point, the service consumer can use the service to utilize a remote service invocation. Cloud broker keeps the location data and available services of all UAV-based fogs to enhance the finding of particular area-based help.

E. Services for Security

Different security mechanisms could be utilized by different UAV-based fogs, IoT devices, and clouds in IoT applications. To regulate security systems among every one of the components is the primary function of the security service in UAV-based fog. And also guarantee that the necessary safety efforts are applied suitably for the protection of the offered services and the physical environment. These services also incorporate authentication, authorization services, access control services for IoT applications, cloud services, and UAV-based fog services. These services can be provided when changing degrees of protection measure such that various applications can utilize the appropriate set of security requirements.

3.5 CHALLENGES FOR UAV-BASED FOG

Key to the situations above is information management and system issues, aside from airborne gadget engineering control, drone, and networking issues. We focus around and summarize difficulties and a scope of issues emerging from managing data in UAV-based fog.

- Planning: as not all user situations can be anticipated, online algorithms are very much needed. In any case, analysis on the historical services exchanges can help to get information about the drones planning. For instance, on the account of the service drones which are used for tourist application, if the requests are coming from a certain tourist spot, or if possibility of delivery where after a few moments, the mapping of frequent routes has to be done, then drones can be repositioned and might be scheduled in view of that.
- Circumstance Awareness: to advise the fog computing infrastructure to help variations to the current circumstance, there is a requirement for setting data. This range from changes in the position of mobile clients to climate changes is being serviced [39]. The UAV drones can move to the positions where the servicing of the customers is expected to require very less energy. A variety of sensors will help to gather the context data and also took care of suitable hubs in airborne fog framework to inform the run-time transformations.
- Drones Utilizing Physical Annotation Information: there is a lot of research for the clients where the visual information can be superimposed on the items seen through a camera by the client, and there is deal with the clients leaving geo-tagged notes for the different clients where clients possibly observe notes when the clients are at the correct areas. Such types of geo-labeled notes can likewise be utilized by the drones for different applications. For instance, one marked an area as "where I had proposed to my husband", in the event and inform robot to travel to that particular place, the framework can resolve to understand state into the GPS organizes and direct robot to move to that particular area. One layer of the semantic names can overlay territories which can be further utilized by the drones for the information about places.

- Reliability: a central point of reliability of the system particularly when basic information should be taken care of. To meet reliability prerequisites in airborne hubs where the numerous things could turn out badly, Redundancy might be a typical solution. It is also possible that a lot of replication of the information is required, which can expand costs. What could be the rate of design consideration for any airborne fog computing foundation which supports an application, where and how such information replication occurs? It will be helpful if the information is divided into grades of required reliability.
- Big-Data from Applications of Airborne Fog Computing: in early 2009, US military robots alone (e.g., thousands working in Afghanistan and Iran) produced a 24 years video whenever viewed it continuously. Along with the utilization of the robots in civilian's applications and services as referenced, an enormous volume of the streaming information will be normal. The management of such information is must, including analysis, preparing, and labeling. There are some other types of difficulties infusing lots of video information for end clients, ground infrastructure in an effective way, and utilizing airborne.
- Drone-Client Interaction: while automation has been enabled in the IoT, there are still a lot of issues where human beings can fit in, like the human in the loop issue; a comparative issue can be seen where autonomous robots should share the control to people, so that the people are not troubled by control drones, particularly. However, drones must have the option to be programmable by clients at a high level. Programming of robots to perform a specific assignment is nontrivial, particularly if the end clients need to do so, such as training robots to take care of specific responsibilities, rather than just predefined tasks. While in brain-controlled drones the exciting trials have been newsworthy, proper abstraction to be associated with the robots will be valuable, for example, train a robot to follow somebody in a specific way or patrol just at specific occasions and places.
- Highways All Around in Air: it is hard to think about robots flying heedlessly in the metropolitan sky; yet to move inside any specifically approved path or "virtual passages" in the sky, the drones can be programmed and additionally underground. An organization of such types of drones, joined by an organization of base stations, of various sizes and shapes, fit to their errands may be utilized; however, the key improvements is expected to execute such type of virtual tunneling.
- Adaptability, Incremental Extensibility, or Compositionality of Sever and Services: service composition to deal with the data arise by such applications will be required, including service for information storage, caching, analytics and reasoning, and handling, and composed on request as indicated by current need. The main thought is that various types of nodes may be associated with various processing stages. For instance, consider any situation where a set of drones wandering on a zone for video capturing the region from alternate points of view. The information may then be streamed to the nearby server, noticeable all around and then relayed

to the ground Cloudlet workers (or sent straightforwardly to the ground servers when reachable). It may be archived on any remote Cloud server in the long run. Video transfers can likewise arrive from the ground robots and cell phones conveyed by clients in the field. The video synthesis, analysis, aggregation, and processing can happen on the servers, which are intermediate. An appropriate systems network capacity and nodes configuration in an integrated architecture that enhances energy and time utilization yet satisfying the application requirements are very much required, and it is anticipated to differ with region and application particular requirement. Those types of services can be compositional and versatile. Over the long duration, if extra types of resources (like more robots or powerful drones) can be added to the system, no significant changes are required. On the other hand if applications build in the incremental method in a steady style (e.g., one drone at an at once), scaled up or out as required, the airborne fog computing system would then need to be adaptable and compositional.

- Communication with Data Control Services for Clients: the client may need to capture video stream as well as view those videos on cell phones after stream and also maybe they can even edit recordings on cell phones. Thus, the drones which can catch those types of video may transfer information to the servers; however, after that an instrument for the clients to improve, alter, or utilize such types of information may likewise be required. Basic solutions exist where clients may very well access their data repositories through the web once they get back; however, an inquiry like which is more adaptable service could be given, like the client can get to the information on request even in remote zone, and even it can control when or where the video is captured by drones which are helping them.

- Optimization for Numerous Factors: a key issue in the above situations is optimizing the energy utilization in any airborne fog processing architectural during the application requirement for the Quality of Experience and QoS. For instance, what is the setup of a drone system for a given fixed budget, airborne fog architecture, and ground-based servers which gives best methods for dealing with video information for the video catching application of tourist in Melbourne, Australia? An organization deciding to make such robot-based video capturing services may utilize the cloud providers which already exist like its own drones and fog computing server. The data mule applications that are utilizing robots may need to exchange time limits for information collection and drone storage limits. One related inquiry is that information transmission and interleaving of thedata processing (like making summaries) would fit the application best.

3.6 APPLICATIONS AND SCOPE OF UAV-BASED FOGS

This part examines various situations where UAV-based fogs can be utilized adequately to fathom a few issues in enormous scope IoT applications.

A. Military Operation

In wars, there are many different types of equipment, structures, vehicles, and correspondence systems. Types of equipment could be radars, sensors, and drones [40]. The military vehicle can be a tank, defensively covered vehicle, monitored or automated airplane, any military ship or boats, and any underwater vehicle. These types of vehicles equipped with the various devices which include the GPS and other special communicating devices which are able to exchange the data among them. The war room could support the advanced automated decisions supportive networks that give the coordination capacities, possible direction, and evaluation for another activity. Also the data about the enemy resource and developments should be gathered and after that it should be analyzed. By keeping that in context, quick choices for action and responses to attack must be taken. UAV-based fogs can give incredible stages to empower secure and effective correspondence and control service for the dynamic armed organizations. The primary advantage is UAV-based fogs' dynamic powerful nature, which permits to discover and associate with particular zones requiring backing and then rapidly moving to help different regions as required.

B. Underwater Observation

The different types of sensors, wireless sensor organizations, or the IoT can be introduced underwater for monitoring various conditions and infrastructure for industrial and security applications [40]. Instances of this scientific application are to observe the underwater climate, for example, geological checking of seas, river floors, oceans, saltiness, temperature, oxygen level, pressure, and bacteria. Another arrangement of the scientific sensor application is utilized to screen ocean creature life, for example, microorganisms, warm blooded animals, and fish. The industrial applications including gas, monitoring the oil-related underwater equipment, and submerged mechanical robots or pipelines [41]. The security application includes observing port offices and coastal borders. Military applications include observing enemy ships and submarine developments. Acoustic communication is normally utilized for underwater correspondence, while the surface floats or station outfitted utilized satellite, WiMAX and GSM for communication. A UAV-based fog can be utilized to encourage underwater checking measures for various applications. UAV-based fog can also replace defective stations to manage underwater sensors, gear, and networks and to add additional abilities and highlights for underwater checking. It could give control and low latency coordination to sensors, gear, or networks [42]. At least one of the UAV-based fogs could be sent to gather data from the underwater sensor and hardware, to launch guidelines and the programmable code to the underwater sensor or gear, also to give real-time service to hardware, and organizations, for example, calculation, communication, or the coordination with control station service.

C. Long Pipeline Monitoring

Long pipelines reach out for many miles in unattended regions, for example, deserts and forest to move gas, oil, or water [43]. It's critical to monitor and control the pipeline structure's condition and transport cycle of the gas, oil, or water in those pipelines. That checking is generally completed by putting various actuators and sensors along the pipeline and interfacing them to the control station which is responsible for controlling, observing, and safety measures. These sensors give the monitoring capacities, including the pressure, temperature, flow rate, and other types of detected information for security purposes utilizing the device, such as camera and motion indicator. Actuators are utilized for controlling pipeline parts, for example, pumps, joints, and valves. The actuators and sensors are generally connected by utilizing the wireless multi-hop networks, wired, or any combination of the both. Many organization faults could cause the disconnected types of fragments on pipeline [44,45]. The fragment can be disconnected from the primary control station and the actuators and sensors on that particular section can't trade any message with main control station.

D. Emergency Response in Large-Scale

In a disaster, volcanoes, earthquake, and huge scale terrorist attack, correct and quick moves should be made within a couple of moments to react and help the influence. In those situations, the different types of IoT hardware could be used which can help in these types of efforts [46]. Wireless sensor networks and sensors could be randomly sent to monitor and help the circumstance. Other types of the UAVs could be additionally used to give quick perceptions for the situation. The ground robot could be utilized in rescue activity. UAV-based fogs could be conveyed to offer service for robots, emergency vehicles, and sensors, and the other UAVs can be used for monitoring purposes.

3.7 TECHNIQUES FOR IMPLEMENTATION AND EXPERIMENTS

A UAV-based fog model was implemented utilizing an appropriated Java agent environment [47,48]. The environment gives a foundation that enables the implementation of advanced platform services for the distributed heterogeneous environment. These types of environments might be consisting of various heterogeneous types of nodes with various capabilities. These middleware framework gives scalable answers to convey, monitor, operate, and control distributed applications. Every specialist in the environment could give run-time backing to the safe execution of code. Those make this framework entirely reasonable to build the advanced platform service to coordinate the cloud with its resources with at least any UAV-based fogs that normally have restricted resources. To implement the model, the greater parts of the core structure of the UAV-based fog stage referenced in Section 3.6 were implemented. These incorporate area-based services and the broker services. A system is additionally implemented to permit a local UAV-based fog brokers to advance a service request to different brokers in the event. Both of the remote and local services invocation were implemented additionally and added to model implementation. We can also utilize the Arduino board [49] that is also open source hardware for the embedded system, for the IoT side. The Arduino was utilized like the IoT payload subsystem which is locally available device requesting services, for implementation of this model.

There are only few sensors which connect with the Arduino, for example, DHT11 sensor [50] for the humidity or temperature estimations. Besides, a few light-emitting diodes (LEDs) and a buzz were introduced to represent actuators. Furthermore, we introduced an Adafruit CC3000 Wi-Fi board [51] to connect the Arduino with a local area network that has a UAV-based fog. The Arduino code was created utilizing the Arduino integrated development environment (IDE) [52] with the Adafruit CC3000 library [53]. Each of the IoT service was implemented using Representational State Transfer Application Programming Interface (RESTful API). At the UAV-based fog side, there is a service that represents every sensor or actuator which is attached to the Arduino. One function of these services is to map a call from the RESTful APIs to Simple Object Access Protocol Application Programming Interfaces (SOAP APIs). All sensor and actuator services are enrolled with the local broker. Furthermore, the global broker is occasionally updated with these services.

3.8 CONCLUSION

Fog computing is empowering technology that can be used for operating and developing IoT-based applications. While the fogs can give low latency services, on the other hand the cloud gives huge scope, resources, and versatile services to IoT applications. Some IoT applications' access in challenging or remote zones is very difficult; they are hard to reach out, for example, forests, mountains, or the underwater. There are lots of models for these types of applications like a pipeline observing and controlling system, underwater monitoring systems, huge scope of emergency response support, and military operations. We introduce a UAV-based fog computing in the chapter. The proposed representation depends upon the fog computing standards. It has additional advantage that the UAV mobility could convey fog to any area wherever it is required mainly. The UAV-based fog offers numerous advantages which are discussed in this chapter. A service-oriented platform for UAV-based fog is also introduced by us. The UAV-based fogs and all IoT resources are seen as the many services that can be utilized to develop any IoT application in this platform. The chapter's contributions are that it can help the researchers understand the basic concepts of UAV-based fog computing and motivate them to implement the new model to assist missions further.

REFERENCES

1. S. Das, R. S. Raw, and I. Das, "Performance analysis of Ad hoc routing protocols in city scenario for VANET," *2nd International Conference on Methods and Models in Science and Technology (ICM2ST-11)*, AIP publishing, New York, pp. 257–261, 2011.
2. R. S. Raw, M. Kumar, and N. Singh, "Security challenges, issues and their solutions for VANET," *The International Journal of Network Security & Its Applications*, vol. 5, no. 5, pp. 95–105, September 2013.
3. A. Hassan, O. Kaiwartya, A. Abdullah, D. K. Sheet, and R. S. Raw, "Inter vehicle distance based connectivity aware routing in vehicular adhoc networks," *Wireless Personal Communications*, vol. 98, pp. 33–54, 2018.
4. R. S. Raw, S. Das, N. Singh, S. Kumar, and Sh. Kumar, "Feasibility evaluation of VANET using Directional-Location Aided Routing (D-LAR) protocol," *The International Journal of Network Security & Its Applications*, vol. 9, no. 5, pp. 404–409, 2012.

5. R. S. Raw, L. A. Kumar, A. Kadam, and N. Singh, "Analysis of message propagation for intelligent disaster management through vehicular cloud network," ICTCS'16, March 05, pp. 1–5, 2016.
6. S. Kumar, A. Bansal, and R. S. Raw, "Health monitoring planning for on-board ships through flying ad hoc network," *Advanced Computing and Intelligent Engineering*, vol. 1089, pp. 391–402, 2020.
7. D. Giordan, A. Manconi, F. Remondino, and F. Nex, "Use of unmanned aerial vehicles in monitoring application and management of natural hazards," *Geomatics, Natural Hazards and Risk*, vol. 8, no. 1, pp. 1–4, 2017.
8. M. F. Pinto, A. G. Melo, A. L. M. Marcato, and C. Urdiales, "Case-based reasoning approach applied to surveillance system using an autonomous unmanned aerial vehicle," in *Proceedings of the 26th IEEE International Symposium on Industrial Electronics*, ISIE 2017, IEEE, Edinburgh, pp. 1324–1329, June 2017.
9. E. Casella, A. Rovere, A. Pedroncini et al., "Drones as tools for monitoring beach topography changes in the Ligurian Sea (NW Mediterranean)," *Geo-Marine Letters*, vol. 36, no. 2, pp. 151–163, 2016.
10. V. Dias, R. Moreira, W. Meira, and D. Guedes, "Diagnosing performance bottlenecks in massive data parallel programs," in *Proceedings of the 16th IEEE/ACM International Symposium on Cluster, Cloud, and Grid Computing*, CCGrid 2016, Colombia, pp. 273–276, May 2016.
11. A. Trotta, M. D. Felice, F. Montori, K. R. Chowdhury, and L. Bononi, "Joint coverage, connectivity, and charging strategies for distributed UAV networks," *IEEE Transactions on Robotics*, vol. 34, no. 4, pp. 883–900, 2018.
12. R. Mahmud, and R. Buyya, "Fog computing: a taxonomy survey and future directions," http://arxiv.org/abs/1611.05539.
13. S. Kumar, A. Bansal, and R. S. Raw, "Analysis of effective routing protocols for flying Ad-hoc networks," *International Journal of Smart Vehicles and Smart Transportation*, vol. 3, pp. 1–18, 2020.
14. S. Kumar, R. S. Raw, and A. Bansal, "Minimize the routing overhead through 3D cone shaped location-aided routing protocol for FANETs," *International Journal of Information Technology*, vol. 13, pp. 89–95, October 10, 2020.
15. B. Schlotfeldt, D. Thakur, N. Atanasov, V. Kumar, and G. J. Pappas, "Anytime planning for decentralized multirobot active information gathering," *IEEE Robotics and Automation Letters*, vol. 3, no. 2, pp. 1025–1032, 2018.
16. D. Erdos, A. Erdos, and S. E. Watkins, "An experimental UAV system for search and rescue challenge," *IEEE Aerospace and Electronic Systems Magazine*, vol. 28, no. 5, pp. 32–37, 2013.
17. S. W. Loke, "The internet of flying-things: opportunities and challenges with airborne fog computing and mobile cloud in the clouds," 2015, http://arxiv.org/abs/1507.04492.
18. C. C. Byers, "Architectural imperatives for fog computing: use cases, requirements, and architectural techniques for fog-enabled IoT networks," *IEEE Communications Magazine*, vol. 55, no. 8, pp. 14–20, 2017.
19. A. Aliyu, A. H. Abdullah, O. Kaiwartya, Y. Cao, M. J. Usman, S. Kumar, D. K. Lobiyal, and R. S. Raw, "Cloud computing in VANETs: architecture, taxonomy, and challenges," *IETE Technical Review*, vol. 35, no. 5, pp. 523–547, 2017.
20. A. Husain, R. S. Raw, B. Kumar, and A. Doegar, "Performance comparison of topology and position based routing protocols in vehicular network environments," *International Journal of Wireless & Mobile Networks*, vol. 3, pp. 289–303, August 2011.
21. F. Bonomi, R. Milito, J. Zhu, and S. Addepalli, "Fog computing and its role in the internet of things," *Proceedings of the First Edition of the MCC Workshop on Mobile Cloud Computing*, ACM, AIRCC, Tamil Nadu, pp. 13–16, 2012.

22. S. Yi, C. Li, and Q. Li, "A survey of fog computing: concepts, applications and issues," *Proceedings of the 2015 Workshop on Mobile Big Data*, ACM, pp. 37–42, 2015.
23. I. Stojmenovic and S. Wen, "The fog computing paradigm: Scenarios and security issues," *Federated Conference on Computer Science and Information Systems (FedCSIS)*, IEEE, Warsaw, 2014.
24. K. Hong, D. Lillethun, U. Ramachandran, B. Ottenwälder, and B. Koldehofe, "Mobile fog: A programming model for large-scale applications on the internet of things," *Proceedings of the 2nd ACM SIGCOMM Workshop on Mobile Cloud Computing*, ACM, pp. 15–20, 2013.
25. M. Satyanarayanan, P. Bahl, R. Caceres, and N. Davies, "The case for vm-based cloudlets in mobile computing," *Pervasive Computing*, vol. 8, no. 4, pp. 14–23, 2009.
26. A. Dasgupta, D. F. Willis, and S. Banerjee, "Paradrop: a multi-tenant platform for dynamically installed third party services on home gateways," *ACM SIGCOMM Workshop on Distributed Cloud Computing*, 2014.
27. J. Zhu, D.S. Chan, M.S. Prabhu, P. Natarajan, H. Hu, and F. Bonomi, "Improving web sites performance using edge servers in fog computing architecture," *IEEE 7th International Symposium on Service Oriented System Engineering (SOSE)*, IEEE, San Francisco, CA, pp. 320–323, 2013.
28. K. Ha, Z. Chen, W. Hu, W. Richter, P. Pillai, and M. Satyanarayanan, "Towards wearable cognitive assistance," *Mobisys*, ACM, ACM, Hampshire, 2014.
29. Y. Cao, P. Hou, D. Brown, J. Wang, and S. Chen, "Distributed analytics and edge intelligence: Pervasive health monitoring at the era of fog computing," *Proceedings of the Workshop on Mobile Big Data*, ACM, ACM, Hangzhou, 2015.
30. M. A. Hassan, M. Xiao, Q. Wei, and S. Chen, "Help your mobile applications with fog computing," *Fog Networking for 5G and IoT Workshop*, IEEE, Seattle, WA, 2015.
31. C. Dsouza, G.J. Ahn, and M. Taguinod, "Policy-driven security management for fog computing: Preliminary framework and a case study," in *IEEE 15th International Conference on Information Reuse and Integration (IRI)*, IEEE, Redwood City, CA, pp. 16–23, 2014.
32. I. Stojmenovic, S. Wen, X. Huang, and H. Luan, "An overview of fog computing and its security issues," *Concurrency and Computation: Practice and Experience*, vol. 28, no. 10, pp. 2991–3005, 2015.
33. F. Bonomi, R. Milito, J. Zhu, and S. Addepalli, "Fog computing and its role in the internet of things," in *Proceedings of the First Edition of the MCC Workshop on Mobile Cloud Computing*, ACM, Helsinki, pp. 13–16, 2012.
34. N. Mohamed, J. Al-Jaroodi, I. Jawhar, H. Noura, and S. Mahmoud, "UAVFog: A UAV-based fog computing for Internet of Things," *IEEE SmartWorld, Ubiquitous Intelligence & Computing, Advanced& Trusted Computed, Scalable Computing & Communications, Cloud & Big Data Computing, Internet of People and Smart City Innovation*, IEEE, San Francisco, CA, pp. 1–8, 2017.
35. Y. Zhou, N. Cheng, N. Lu, and X. S. Shen, "Multi-UAV-aided networks: Aerial-ground cooperative vehicular networking architecture," *IEEE Vehicular Technology Magazine*, vol. 10, no. 4, pp. 36–44, 2015.
36. K. Bilal, and A. Erbad, "Impact of multiple video representations in live streaming: A cost, bandwidth, and QoE analysis," in *Proceedings of the 2017 IEEE International Conference on Cloud Engineering*, IC2E 2017, Canada, pp. 88–94, April 2017.
37. Open Connectivity Foundation Website, https://openconnectivity.org/, viewed March 24, 2017.
38. Z. Chen, F. Xia, T. Huang, F. Bu, and H. Wang, "A localization method for the Internet of Things," *The Journal of Supercomputing*, vol. 63, pp. 1–18, 2013.
39. Y. Liu, J. E. Fieldsend, and G. Min, "A framework of fog computing: Architecture, challenges, and optimization," *IEEE Access*, vol. 5, pp. 25445–25454, 2017.

40. D. Zheng and W.A. Carter, *Leveraging the Internet of Things for a More Efficient and Effective Military*, Rowman & Littlefield, Lanham, MD, 2015.
41. M.C. Domingo, "An overview of the internet of underwater things," *Journal of Network and Computer Applications*, vol. 35 no. 6, pp. 1879–1890, 2012.
42. P. Houze, E. Mory, G. Texier, and G. Simon, "Applicativelayer multipath for low-latency adaptive live streaming," in *Proceedings of the ICC 2016-2016 IEEE International Conference on Communications*, Kuala Lumpur, Malaysia, pp. 1–7, May 2016.
43. N. Mohamed and I. Jawhar, "A fault-tolerant wired/wireless sensor network architecture for monitoring pipeline infrastructures," in *Proceedings 2nd International Conference on Sensor Technologies and Applications (SENSORCOMM 2008)*, IEEE Computer Society Press, IEEE, Cap Esterel, pp. 179–184, August 2008.
44. N. Mohamed, J. Al-Jaroodi, and I. Jawhar, "Modeling the performance of faulty linear wireless sensor networks," *The Int'l Journal of Distributed Sensor Networks – Special Issue on Verification and Validation of the Performance of WSN*, vol. 2014, Article ID 835473, 12 pages, 2014.
45. N. Mohamed, J. Al-Jaroodi, I. Jawhar, and S. Lazarova-Molnar, "Failure impact on coverage in linear wireless sensor networks," in *Proceedings Int'l Symposium on Performance Evaluation of Computer and Telecommunication Systems (SPECTS 2013)*, IEEE Communications, IEEE, Toronto, ON, pp. 188–195, 2013.
46. L. Yang, S.H. Yang, and L. Plotnick, "How the internet of things technology enhances emergency response operations," *Technological Forecasting and Social Change*, vol. 80 no. 9, pp. 1854–1867, 2013.
47. J. Al-Jaroodi, N. Mohamed, H. Jiang, and D. Swanson, "Middleware infrastructure for parallel and distributed programming models on heterogeneous systems," *IEEE Transactions on Parallel and Distributed Systems, Special Issue on Middleware Infrastructures*, vol. 14, no. 11, pp. 1100–1111, November 2003.
48. J. Al-Jaroodi, N. Mohamed, H. Jiang, and D. Swanson, "An agent-based Infrastructure for Parallel Java on Heterogeneous Clusters," in *Proceedings 4th IEEE Int'l Conference on Cluster Computing (CLUSTER 2002)*, IEEE, Chicago, IL, pp. 19–27, September 2002.
49. Arduino website, https://www.arduino.cc/, viewed March 24, 2017.
50. DHT Sensor Library Website, https://github.com/adafruit/DHTsensor-library, viewed March 24, 2017.
51. CC3000 Wi-Fi board Website, https://www.adafruit.com/products/1469, viewed March 24, 2017.
52. Arduino IDE Website, http://arduino.cc/en/main/software, viewed March 24, 2017.
53. Adafruit CC3000 Library Website, https://github.com/adafruit/Adafruit_CC3000_Library, viewed March 24, 2017.

4 Wi-Fi Computing Network Empowers Wi-Fi Electrical Power Network

Y.P. Chawla
Australian Graduate School of Leadership

CONTENTS

4.1 Objectives of the Chapter 51
4.2 Increased Flexibility a Must for the Future Power Utility Constructs 51
4.3 Energy Importance for Data Centres and Network Stations and
 Costs of Energy 53
4.4 Computing Has Full Synergy with Energy 54
4.5 Wireless Power Transmission 55
4.6 Leadership in Innovation 60
4.7 Long and Short of Wi-Fi 61
4.8 Conclusions 62
Acknowledgements 63
Bibliography 63

Background: For transforming our world by adopting the UN's Sustainable Development Goals (SDG-2030), adapted by 193 member state's signatories. UN Sustainable Development Goal-7 focuses on various stakeholders' efforts to ensure trifecta parameters of reliable, affordable, sustainable power to all. SDG-7 requires ensuring cleaner energy deployment to reach the remotest place of living globally by any human being. The cost of power transmission per head becomes uneconomical with sparingly located populations at remote locations. Traditionally, electrical conductors help electricity transmission. Higher voltages in transmission lines helped achieve higher transmission efficiencies. The introduction of Wireless Power Transmission (WPT) and Virtual Power Transmission (VPT) saves on the cost of electrical conductors and transmission towers, making it economical to transmit power.

Emrod, a New Zealand (NZ) company, has created a sensation by demonstrating the first iteration of a commercial WPT system. One of the NZ's largest power utilities supports Emrod in this development phase, with a view to then implement WPT within their network to address issues with power transmission in rugged terrains or remote locations (Blain, [5]).

The grid operator (Transmission company) Re'seau de Transport d'E'lectricite' (RTE) in France has deployed a 12MW/24 MWh battery storage system. To ensure market neutrality and release the stored electricity through virtual transmission, the surplus power generated and not immediately needed for consumption is stored in a battery system. Thus, the battery systems avoid constructing new power lines to add in electricity at the time of peak loads. The computing network triggers the flow of stored power into the transmission network when required. This mode of power transmission is called Virtual Power Transmission (VPT).

WPT helps transfer power from one geography to another not connected on a power grid. At the same time, the VPT uses excess energy not required earlier and stored. Thus avoiding the flow of power at peak hours and avoiding transmission congestion requiring additional investment. Transmission of the stored power during non-transmission congestion time saves on enhancing or deferring the grid capacity augmentation investments.

The Chapter's Mission: Mission is to make a productive partnership and create a force multiplier of computer science experts and the power sector experts or the budding experts appreciative of inputs to their respective field of expertise and knowledge. This chapter shares knowledge about one of the significant inputs to the computer data centres, i.e. electrical power, to understand the cross-disciplinary subject. Over the years, the generation of power and energy storage has progressed, and ways are evolving to improve its optimum utilisation. The electrical power transmission achieved efficiencies by going in for higher voltages of ac or dc transmission. It intends to enhance the computing enhance collaborative approach in computing and electrical power, enhancing **Computing synergy with energy**.

Concerns About the Subject: WPT has resurfaced after 130 years. With the development work continues in NZ and upcoming field trials in NZ and other countries, it seems commercial WPT realisation is imminent. The project's success opens the potential for cost savings in electrical power transmission, especially installing traditional poles and wires is challenging. Similarly, VPT has already established its credentials in Australia, France and the USA. India is now working on the concept. While this chapter highlights the power sector association and its dependence on computing, the cyber attacks on the power grid via computing software systems alert from getting vulnerable due to such cyber attacks (*Critical Infrastructure Protection II, eds*, 2008 by *International Federation for Information Processing*).

Wireless Power Transfer is getting deployed in many applications other than power sector and is segmented as under[1]:

i. Technology Segmentation: Inductive, Magnetic resonance, conductive. Radio Frequency and Infrared
ii. Segmented by Range of Transmission: Near Field and Far Field
iii. Segmented by Applications: Power, Consumer Electronics, Automotive, Industrial, Health care, Defence, Others
iv. Segmented by Geography: North America, Latin America, Asia Pacific, Europe, and the Middle East and Africa as voltage levels are different.

---

[1] https://www.wboc.com/story/43376858/wireless-power-transmission-technology-market-share-size-global-future-trend-segmentation-business-growth-top-key-players-analysis-industry; https://www.fortunebusinessinsights.com/industry-reports/wireless-power-transmission-market-100567

Research on Wire-free power transfer started in the 1880s and slowly dragged on and picked up again in the 2010s. Post COVID pandemic, when everyone started looking for green energy, the momentum on WPT heated up (Anon [3]).

All these applications are not possible without the support of Wi-Fi in the local network. The Wire-free makes the system worry-free and easy to operate. Wi-Fi computing is powering the machines directly from the remote. The computing experts are required to be ready for the future and create a new future.

4.1 OBJECTIVES OF THE CHAPTER

"Leadership in Innovation" in the computing-related domain triggered the concept of this chapter. Introductory remarks on leadership in innovation have highlighted. India is working on One Sun, One World, One Grid. All these applications need massive support of computing.

  i. The COVID-19 has changed our future of work, with more organisations adopting Cloud Computing, the Internet of Things and other technologies supported by computing. Considering a renewed thrust on green energy generation as a post-COVID impact, a new approach to working in the power sector and computing collaboration is an objective. India is working on the same.
  ii. The global electricity consumption as an input to computing data centres ranges from 1% to 2% of global power generation. (Variation is due to results of the various studies.) The cost of electricity can change the bottom line of computing as an industry. Examining the impact of wireless electricity power consumption on computing technology has been analysed.
  iii. Computing Scientists in the hardware industry are focused on making the systems energy efficient, developing the hardware for networks, ad-hoc networks: wireless sensor networks (WSN), wireless mesh networks (WMN), vehicular ad-hoc networks (VANET) and the like, examining the collaborative approach for futuristic technology as in the subject.
  iv. Moreover, software scientists and engineers are working hard on changing the characteristics of any helpful information to the Artificial Intelligence (AI) data. The AI support for oncoming technology is analysed.

4.2 INCREASED FLEXIBILITY A MUST FOR THE FUTURE POWER UTILITY CONSTRUCTS

The power generation in hybrid mode comprising thermal and infirm renewable (solar and wind) are getting smarter and ramping up or down as fast as possible. Power generation smartness is required to meet the changing load requirements. Work on the smarter power generation is already progressing. Computing hardware and software are aiding the power generation and maintenance of the power plants. The advanced algorithms help identify and predict the equipment problems better than most of the team members involved in the operation power plant. Smart sensors used in ad-hoc networks help in computing for such predictions on installed explicitly for such purposes. The computing systems analyse the power system's real-time data through the Cloud and have a built-in analysis of the data.

Similarly, on the demand side of power management, the distribution networks are called smart grids. The digital twins comprising the physical distribution network and the electricity content flow that bridges the distribution network with the computer network have been in operation for some time now. The computer system collects the real-time data of the ambient temperature, time of the day and the distribution grid's power requirements by analysing, manipulating or optimising its power requirements. Over the years, terms used include hybrid twin technology, "virtual-twin" and digital asset management. Alteration in the control logic of the digital twin can help analyse the digital power plant's response. The same, if found acceptable, can go in for a real-time change in resolving the control problem. Thus, increased reliability, assisting in knowledge transfer, and most important is allowing innovation.

Somehow, the electricity transmission system has yet to be made interactive with innovative control centres, intelligent transmission networks and smart sub-stations – power loading or unloading point of connections. Transmission networks are victims of natural calamities like earthquakes, tornados, hurricanes and cyclones, and cyber attacks. The Electrical Transmission network inherits the infrastructure challenges of long transmission lines, inhabitable transmission tower footprints and right of way in the habitable lands. The support for fast online analysing tools, wide-area monitoring, measurement and control of power flow and accurate protections is needed for the network's reliability. The power transmission system needs innovative technologies and advanced power electronics controlled through IoT.

Computing has helped to have grid flexibility in the power system. The stability is required on variable time scales such as short, medium and long terms. The time spans from sub-seconds to a few seconds, from seconds to minutes covering short-term frequency control and even covering minutes to hours for the power system's stability. The fine-tuning is required to meet the supply and demand side fluctuations, days to months or even to years, to cover seasonal or inter-annual power requirements. Computing can thus support the WPT for dynamic-stability inertia response to maintain voltage and frequency, with the extensive data analysis support and help in scheduling energy over long durations.

Analysing the collected data is not new to the power utilities. The computing is undertaking the end-to-end analysis support covering from the consumer meters to distribution and transmission sensors and control devices. Historically this data has covered the distribution sector and now required for WPT and VPT. There are myriad data streams that can extend the benefits of extensive data analysis, waiting for Federal Communication Commission USA to realign its 900 MHz broadband and then allowing a 3.5 MHz broadband spectrum. Further auctioning of the 3.5 MHz as Citizens Broadband Radio Service (CBRS) broadband spectrum is possible. Power utilities are pursuing the deployment of private Long-Term Evolution (LTE) networks in both bands for grid management support. The private broadband network based on the recognised standards helps real-time communications and support to WPT and VPT and the power utility to collect and analyse big data. With high advances in machine learning, AI, Cloud services on hyper-scale and the advent of "data lakes" allow data storage at a large scale, structured or unstructured. The power utilities can undertake advanced analytics for improved safety, efficiency and system security.

The importance of grid data collection has increased WPT and VPT because of system addition and intermittent electricity flow.

4.3 ENERGY IMPORTANCE FOR DATA CENTRES AND NETWORK STATIONS AND COSTS OF ENERGY

The data centres got entrusted with new challenges because of their continuing success. The enhanced requirements are remote weather collection and communicating to a base station. The remote data centres require electricity as input. The border conflicts in inhospitable terrains require video eye and the enemy movement data. These situations require remote data centres for weather data, serving as an eye hawk and wireless power for the data centres and the armed forces' communication as in Figure 4.1.

Further, the recent development of data centres of hyper-scale, Cloud centres and network centres generally has higher energy efficiency than the centres that have come up as a part of the expansion in the organisations' setup. The leaders of such new centres or the hyper-centres understand the value of energy efficiency. Further emphasis requires understanding the impact on the bottom line of the data centre through a lesser input energy cost[2] (Arya College [2]).

These hyper-scalers have turned to green energy like solar and wind. Microsoft stores its data files under the sea to keep it cool for lesser energy consumption to reduce energy costs, keeping it in air-conditioned space.

Computer scientists' involvement in the energy sector is already going on. The hyper-scale data centres, Cloud centres and network centres welcome the advantage of cost reduction. Thus, WPT's achievement is also achievable and mutually advantageous to power and the data centres with their collaborative approach.

For the UN's sustainable development goals for affordable and clean power, identifying the enablers and inhibitors that AI provides to the power sector gives way forward.

 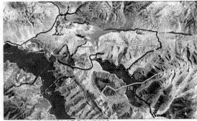

FIGURE 4.1 Terrain of India China Border with no power and difficult communication.

---

[2] https://ieeexplore.ieee.org/document/7153532 & https:// aryacollegejaipur.medium.com/wireless-power-transmission-technology-with-applications-219357ef8aa2#:~:text=WPT%20can%20use%20in%20moving,Rectifying%20Circuits%20and%20Wireless%20sensors.

One such inhibitor is power consumption by the control centres. However, the same (Nicola, 2018) got negated by another study proving that the control centre growth outpaced the energy consumption many folds as the control centres are more energy efficient. The energy delivery at a specific time and to whom to economise the energy delivery cost is AI's capability.

However, "concerns about the subject" at para 2.2 above raises the most concerning aspect of computing and the power sector association:

- Cloud Computing – Energy Efficient: Cloud Computing has made data centres energy efficient. Large companies like Google, Facebook, Amazon and Microsoft own the largest cloud centres, which extend to football-sized fields. India is going in a big way to establish data centres in India. Microsoft data centres have been even storing the data undersea to reduce the power requirements. The computing data centres are self-conscious of energy consumption at their centres. Data centres' growth is sixfold over the years (2010–2018) compared to the energy demand of just a 6% increase. Thus, computing has been doing greater good for humanity while supporting all sectors, including the power sector. The cloud services expansion for additional data analysis can be modular, wherein the power flow is the traffic flow. Computing can help quick and efficient accident detection making the mobile WPT system ready to operate (Lohr).

4.4 COMPUTING HAS FULL SYNERGY WITH ENERGY

Power sector has already reduced energy costs by adding green power replacing conventional power. Solar energy and wind energy are though infirm power, as the generation changes with timing, the month of the year, and similar parameters. Computing has made it possible for power grid synchronisation of infirm power with conventional firm power to enable the power sector to reduce the overall costs per kWh terms.

With continuing computing support to the power sector to help the cost reduction of power on WPT, access to cheaper power is possible. The WPT and VPT can go on together. The VPT, as referred to in Australia and India, is also called grid boosters in Germany. Digitalisation is under immense pressure to meet the challenges of making the power grid smarter. The smartness requirements must meet the transmission grid requirements of infrastructural challenges created by ageing network and insufficient investments, increased power requirements, power transmission from the firm and infirm power generation, virtual power plants, etc. Digitalisation is required to help the power grid manage electricity's economic delivery while reaching the remotest corners and meeting reliability and energy quality.

Further, innovative digital support is required to analyse real-time and predictive modelling along with security analysis.

Digitalisation inputs are well accepted, but the power grid on its side needs innovative technologies, materials, efficient power electronics, data availability for computation and AI, communication and control features of the power grid systems. These innovations have come as WPT and VPT.

4.5 WIRELESS POWER TRANSMISSION

The NZ government, through its innovation institute, is backing commercial scale WPT development. NZ's second largest power distribution utility, with around 1.1 million customers, also provides backing to the same development.

WPT helps transmit electricity to remote areas like front defence posts. Figure 4.1 the Islands of Andaman and Nicobar and the Islands of Lakshadweep in Figures 4.2 and 4.3, respectively.

The islands which are far away from the mainland and unable to be connected to the national grid have been dependent on diesel power generation. Solar power and energy storage battery systems have resolved the issue to some extent. However, having such a system at every island could be a challenge; thus, the power generated on a larger island could be an environmental ramification for the blue economies (ocean economies).

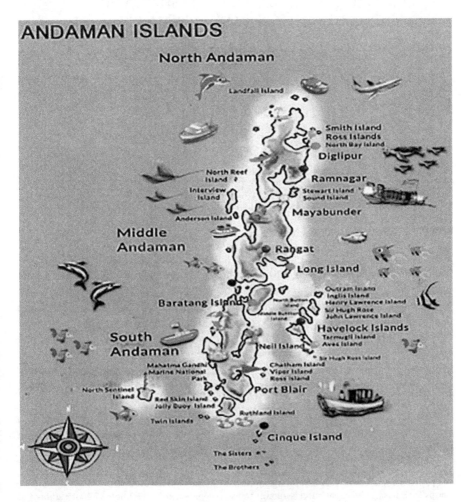

FIGURE 4.2 No electrical power grids in islands – Andaman & Nicobar India.

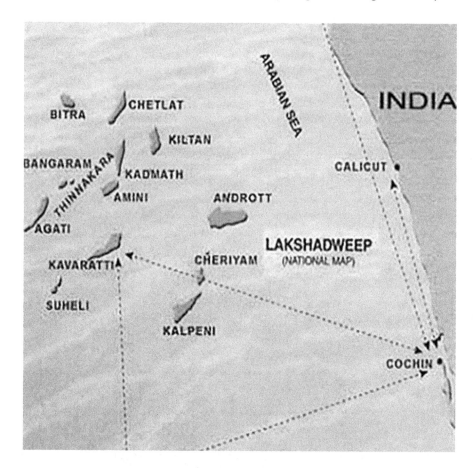

FIGURE 4.3 Challenge of electricity power transfer in islands – Lakshadweep India.

Challenging terrains such as mountainous regions or large water bodies represent significant challenges to installing traditional wires and poles. WPT systems are employable to a substantial advantage in these situations.

Helicopters are assisting in meeting the transmission lines construction challenges in a problematic area where reaching such locations is expansive, and WPT can help.

These problematic areas have much land where renewable energy projects like solar and wind, where installation without any human maintenance is not a problem. The generation of power of such projects by drowns possible due to maintenance. Power transmission in such areas can be by WPT. WPT symbolically has been depicted in Figure 4.4.

Japan[3] and NASA USA[4] are working on wireless power transfer from space to vehicles in space and to the earth from space. The above efficiencies and data, as in Figure 4.5, are from one such research paper (Chaudhary & Kumar, 2018).

---

[3] https://edujournalfuturesreasearch.springeropen.com/ articles/10.1186/s4009-018-019-7
[4] https://sbir.nasa.gov/content/wireless-power-transmission#:~:text=The%20focus%20of%20this%20activity,sites%2C%20between%20ground%20and%20space

Wi-Fi Computing Network

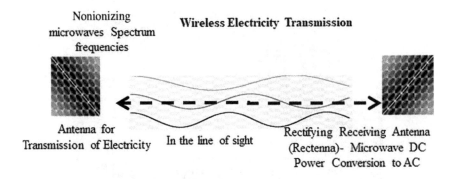

FIGURE 4.4 Wireless Power Transmission – symbolic.

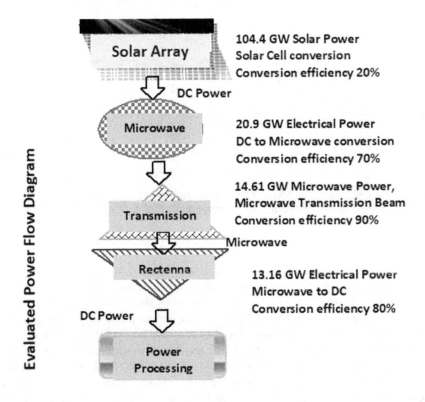

FIGURE 4.5 Wireless power transfer – satellite solar power for ground supply: a futuristic, clean energy system.

The importance of the line of sight is as vital as in the case of a satellite. The view of two antennas in one line is the technical compulsion; the power transmission distance can be 20 m or 20 kms. The radial distance decides the component structure, like the size of the antennas. Figures 4.6 and 4.7 show the antennas for power transmission.

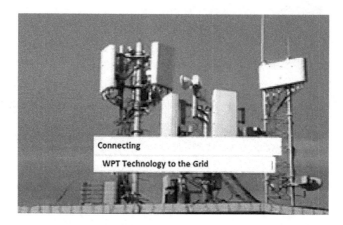

FIGURE 4.6 Wireless Power Transmission – connecting two points in the line of sight.

FIGURE 4.7 Wireless Power Transmission – connecting two points in the line of sight.

Mediating technologies help regain signal fidelity and refocus the power beam of transmission power through the air medium using the rod's relay technology. In WPT, electromagnetic waves[5] over long distances by shaping the beam of the electromagnetic waves, meta-materials, and using rectenna technology. Going from a wired system to wireless on many stretches can save on material costs and faster implementation. The project uses solid-state electronics as in defence radars. Further developments are evolving due to computing and the 5G Wi-Fi. The tower structure is more straightforward as compared to the conventional transmission towers unless the antennas and rectennas (receiving antennas) are installed on the existing transmission tower to add to the transmission capacity.

The structures of the WPT are simpler, as in Figure 4.8. WPT can also support the powering of drones,[6] as in Figure 4.9 for a tireless drones duty.

---

[5] https://new-zealand.globalfinder.org/company_page/emrod#lg=1&slide=2; https://spectrum.ieee.org/energywise/energy/the-smarter-grid-emrod-chases-the-dream-of-utilityscale-wireless-power-transmission

[6] https://new-zealand.globalfinder.org/image_cloud/136c0a4a-ab94-11ea-bc6f-1dcf19923f8d/image?full=1.

FIGURE 4.8 Structure to mount the WPT.

FIGURE 4.9 Powering the drones on extensive surveillance.

Virtual Power Transmission: Using AI and digital intelligence for electrical sector reforming is like topping up century-old grid 1st-century technology on the 20th-century grid as indicated by Fluence (Siemens & AES company).[7] It is a long-term digital strategy and an open ecosystem for the power sector. The VPT has come out as a part of such intelligence and considered as equitable to finding an alternate route to slip through when stuck in a traffic jam. The traffic jam analogy fits very well with the transmission congestion of the power grid. Alternatives are diverting power to the battery and transmitting power when feasible during reduced transmission congestion, similar to redirecting the vehicular traffic through other lanes or routes that come out from thin air when additional flyovers or underpasses are not available.

---

[7] https://blog.fluenceenergy.com/fluence-ai-and-digital-intelligence-drive-electric-transformation.

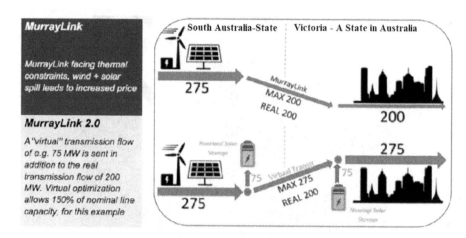

FIGURE 4.10 Transmission line capacity 200 MW, additional 75 MW transferred and stored in the battery avoiding line congestion- symbolic layout.

Figure 4.10 depicts an actual situation of virtual transmission in South Australia to Victoria. The transmission system's power-carrying capacity is 200 MW at peak hours against the power requirements at 275 MW in Victoria. The addition of 75 MW is causing transmission congestion towards Victoria. Alternate to making additional investments in Grid infrastructure expansion, the battery system at dispatch and receiver end installed can help release the stored power from South Africa to Victoria during non-peak hours and hold the energy on Victoria side. The stored energy is transmitted when needed to meet the peak load. This system requires in-depth analysis and computing switches. Figure 4.10 tells the story.

Thus, the VPT acts as a redrawn grid network map without additional grid support investments. The Australian government has acknowledged the virtual power and agreed to support the efficient transmission network augmentation by a non-network solution like virtual transmission lines.[8]

The integrated hardware, digital intelligence and the required software can work out the best business case to achieve the optimal energy storage value. The stored energy is virtually transmitting the same or WPT. The process supported by AI, advanced forecasting and power portfolio optimisation based on the marginal cost of power transmission through WPT or VPT (out of stored energy) compared with the conventional power grid's operating cost.

4.6 LEADERSHIP IN INNOVATION

The innovative concept, as indicated above, is now ready for commercial launch. The knowledge share of the inventive concept of the Wi-Fi transmittal of electric power having crossed the pilot stage and is on the way to commercial launch is the focus of this chapter. It requires the following:

---

[8] https://www.energymagazine.com.au/aemo-outlines-west-murray-zone-challenges/.

i. The collaborative approach already developed between computing and electrical power needs to work further to reduce electricity costs and access.
ii. Transmitting electricity generated in remote distribute locations to be aggregated and transmitted wirelessly to places of consumption.

Lots of ideas are with the managers associated with computing or the power sector. The managers in these sectors are adequately informed and know theories with abundant knowledge and expertise. These managers are required to work with innovative thinking beyond the traditional business. The pause during the COVID-19 pandemic gave us time to restart with creative thinking for leadership thoughts. This leadership requires innovation rather than managing the innovation activities creating an environment for innovation where all the team members think of the invention of a specific concept or product.

A leader's vision captures the futuristic technologies, and an innovative business leader translates it to a business opportunity for oneself, a team of one's organisation. The power sector is developing the innovation to integrate renewable energy with the power grid; computing continuously evolves through its innovative approach. WPT is another creative process that gets suitable technology partners for further growth.[9] Several papers written on the subject concluded that WPT is an excellent futuristic technology. The innovation comes out of readying a rapid prototype. The next step is proceeding to a "try-it-out approach" and gaining hands-on experience for the innovation's invaluable inputs. WPT has already gone beyond the stage of building and testing the latest technology. It has jumped past the information overload on WPT. The WPT's innovation has been possible with a focused team created to aggregate their respective knowledge and worked quickly. Putting together the lessons learnt, the vibrant team proceeded to go in for patenting the technology. The teamwork of the said project development communicated well and accepted each other's ideas, setting up space and defining processes for healthy interaction and ease of exchange of ideas, fun and a serious play as indicated by the team leader. Under innovative leadership, the team leader allows the team members' creative thinking and supports what is better and what next.

Industrial revolutions up to 4.0 marked significant technological development. The WPT can be a part of the 5.0 Industrial revolution, seeking innovation connect to the purpose and inclusive approach to help in a global inequality as a significant underlying cause.

4.7 LONG AND SHORT OF WI-FI

The AI took over the landing of Mars-rover-2021 with terrain relative navigation. Thus the computing has the power for failsafe application of technologies. Communication was not direct with space vehicle, but through NASA's Deep Space Station, an international network of antennas felt safer over a long distance. The data rate natural to earth varies from 500 to 32,000 bits/sec, roughly half the modem's speed at home. Power and communication through Wi-Fi are interdependent, that Mars power conservation is ensured by talking to earth. The Perseverance rover carries radioisotopes that give

---

[9] https://www.ccl.org/wp-content/uploads/2015/04/InnovationLeadership.pdf.

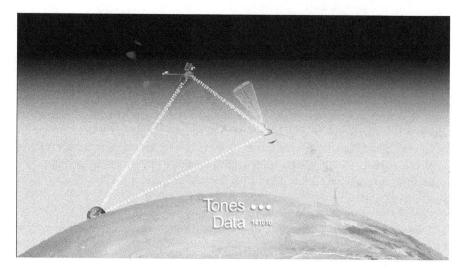

FIGURE 4.11 Mars Perseverance rover communication.

input power for communication signals, and to avoid its yelling, the rovers placed at 257 and 400 km from Mars.[10] Power and Wi-communications are so much interdependent. The interdependency between power and Wi-Fi used in the community battery concept storing solar energy by the community members and application in the virtual power plants applications found usefulness. Figure 4.11 indicates the Mars communications.

4.8 CONCLUSIONS

The WPT and VPT make a perfect case of making Aatmanirbhar Bharat (Self-reliant India) and making the electricity reach everyone globally as per SDGs. The development also fits in as a project under Sanrachana of Bharat Heavy Electricals Limited (BHEL) entrusted by the Ministry of Industry of India's government. The other important stakeholders to whom the WPT is helpful is the electrical and mechanical core of engineers (EME) for our border areas. Ministry of new and renewable energy Govternment of India (GoI) can consider generating renewable energy (RE) in remote areas and transmit the same wirelessly. The power transfer from island to island is another challenge that WPT and VPT can meet.

Without the support of the Ministry of Electronics & Information Technology & Communication – Government of India, things cannot move on this subject. The most important stakeholders are computing and power teaching fraternity. The chapter provides leadership in innovation and the futuristic technologies of power in collaboration with computing.

With WPT and Wi-Fi data support, we can build a global power grid and swap power worldwide. It needs a changing infrastructure for power supply, which has not seen an extensive change in wired power supply technology. The underwater cables and running pylons and wires through harsher terrains has been expensive to maintain.

---

[10] Mars.nasa.govt/msl/mission/communications/#data.

Insights from these innovative efforts drive the futuristic technologies and enhancement of the grid performance. We can build a global power grid that needs DC wizardry[11] and power efforts to reach any part of the world and meet the goals of SDG-7 across the continents. The long-range Wireless Power Transfer through microwaves has shown great potential using Ghz levels of frequency through phased arrays and rectennas. Such a system can be driving power to solar-powered satellites, powering drone aircraft, charging wireless devices in addition to lighting the houses and industry.

It is well established that the future belongs to big data, and the power sector growth depends on utilising private broadband networks. The positioning of both the power sector and the computing industry is such to ride that wave. The SDGs of the UN of removing global inequality are achievable by making the clean, economical and affordable power reach the remotest location.

Can these technologies of WPT, VPT and Virtual Power Plants survive without Big Data Analytics in Wireless Ad-hoc Networks Enabled by Cloud Computing? The answer is a big No. There is enormous scope for the computing fraternity.

ACKNOWLEDGEMENTS

The contribution of **Dr Fayed Ramzi, the Executive Dean of the Australian Graduate School of Leadership**, is thankfully acknowledged having encouraged to touch upon innovation leadership in this innovative high technology project. Inputs provided by Emrod New Zealand on their latest project under the patenting are also gratefully acknowledged.

The editors' proactive approach needs special acknowledgements to have taken up the futuristic technology as a part of the book, making the computing community ready on the subject.

BIBLIOGRAPHY

1. Karabulut, A. et al. (2018, October 24). Wireless power transfer and its usage methods in the applications. *Research Gate*, www.researchgate.net/publication/327981251_Wireless_Power_Transfer_and_its_Usage_Methods_in_the_Applications. Accessed 30 Nov. 2020.
2. Arya College, A. (2020, January 21). Wireless power transmission technology with applications. Aryacollegejaipur.Medium.com; Arya College Jaipur. https://aryacollegejaipur.medium.com/wireless-power-transmission-technology-with-applications-219357ef8aa2#:~:text=WPT%20can%20use%20in%20moving.
3. Anon. (n.d.). *Wireless Power Transmission Market Size, Industry Share And Growth Rate 2019–2026*. www.Fortunebusinessinsights.com. Retrieved 19 Dec. 2020, from https://www.fortunebusinessinsights.com/industry-reports/wireless-power-transmission-market-100567.
4. Birshan, M., Goerg, M., Moore, A., & Parek, E.-J. (2020, October 2). *Investors Remind Business Leaders: Governance Matters | McKinsey*. www.Mckinsey.com; Mckinsey & Company. https://www.mckinsey.com/business-functions/strategy-and-corporate-finance/our-insights/investors-remind-business-leaders-governance-matters.

---

[11] https://en.wikipedia.org/wiki/Wireless_power_transfer.

5. Blain, L. (2020, August 3). NZ to trial world-first commercial long-range, wireless power transmission. New Atlas, New Atlas. Com, newatlas.com/energy/long-range-wireless-power-transmission-new-zealand-emrod/. Accessed 30 Nov. 2020.
6. Chaudhary, K., & Kumar, D. (2018). Satellite solar wireless power transfer for base-load ground supply: clean energy for the future. *European Journal of Futures Research*, 6(1). Doi: 10.1186/s40309-018-0139-7.
7. Dogan, O., & Ozdemir, S. (2020, October 21). A review of wireless power transmission in the applications. *5th International Mediterranean Science and Engineering Congress (IMSEC)*, Antalya, Turkey.
8. Drupal. (2013, October 31). Wireless power transmission. *Small Business Innovation Research NASA; Small Business Innovation Research / Small Business Technology Research.* https://sbir.nasa.gov/content/wireless-power-transmission#:~:text=The%20focus%20of%20this%20activity, sites%2C%20between%20ground%20and%20space.
9. Galura, B., & Madaeni, S. (2020, 15 October). Applying 21st century technology to 20th century grid: using AI and digital intelligence to drive the electric transformation. Blog.Fluenceenergy.com; Fluence - Siemens and AES. https://blog.fluenceenergy.com/fluence-ai-and-digital-intelligence-drive-electric-transformation.
10. Hartmann, I. AEMO outlines west murray zone challenges. *Energy Magazine*, 28 Feb. 2020, www.energymagazine.com.au/aemo-outlines-west-murray-zone-challenges/. Accessed 8 Dec. 2020.
11. Horth, D. (2014). Innovation leadership. *How to use innovation to lead effectively, work collaboratively and deliver results, center for creative leadership.*
12. IEEE Spectrum. (2015, July 28). Let's build a global power grid. *IEEE Spectrum: Technology, Engineering, and Science News*, IEEE, spectrum.ieee.org/energy/the-smarter-grid/lets-build-a-global-power-grid. Accessed 5 Dec. 2020.
13. Irena, International Renewable Energy Agency. Artificial Intelligence and Big Data Innovation Landscape. www.irena.org, 2019.
14. Li, Y. et al. (2020, November 13). Directional characteristics of wireless power transfer via coupled magnetic resonance. *Electronics*, vol. 9, no. 11, p. 1910, Doi: 10.3390/electronics9111910. Accessed 29 Nov. 2020.
15. Lohr, S. Cloud computing is not the energy hog as has been feared. *The New York Times*, 27 Feb. 2020, www.nytimes.com/2020/02/27/technology/cloud-computing-energy-usage.html. Accessed 29 Nov. 2020.
16. New Zealand Global Finder. Wireless power transmission. *Globalfinder.org*, 2020, new-zealand.globalfinder.org/image_cloud/9bb0ce8f-ab94-11ea-9188–916555f46316/image?full=1. Accessed 14 Dec. 2020.
17. ScaleUP NZ. ScaleUP NZ - EMROD. *Scale-Up New Zealand*, Explore NZ Innovation, 2020, new-zealand.globalfinder.org/company_page/emrod#lg=1&slide=2. Accessed 10 Dec. 2020.
18. Wikipedia Contributors. (2019, March 13). Wireless power transfer. *Wikipedia*, Wikimedia Foundation, en.wikipedia.org/wiki/Wireless_power_transfer. Accessed 1 Dec. 2020.
19. Wireless power transmission using technologies other than radio frequency beam SM series spectrum management. international telecommunication union (ITU), Report ITU-R SM.2303-2, 2017.

5 Big Data Analytics for Vehicular Edge Networks

Jayashree Patil
MIT Academy of Engineering Alandi (D)

Suresha
Sri Venkateshwara College of Engineering

Nandini Sidnal
KLEs M.S.Sheshagiri College of Engineering

CONTENTS

5.1	Introduction	65
	5.1.1 Impacts of Intelligent Computing Technologies in VANET	67
	5.1.2 Wireless Communication Technologies	69
5.2	Big Data Analytics	70
	5.2.1 Data Mining Techniques in the VANET	70
	5.2.2 Machine Learning for VANET	72
5.3	Edge-Enabled Data Gathering and Aggregation	73
	5.3.1 Data Gathering	73
	5.3.2 Data Aggregation	74
5.4	Edge-Enabled Service Content Prefetching and Storing	74
5.5	Edge-Enabled Computing	75
5.6	Result and Discussion	76
5.7	Data Analysis	78
5.8	Conclusion	80
References		81

5.1 INTRODUCTION

Traffic congestion represents a serious issue with urbanisation growth and an increased number of vehicles (Liu et al. 2017). The transportation system needs to redefine the traffic policy to save life on roads (Contreras et al. 2017). The primary reason for road congestion is that at peak time at particular roads, vehicle density increases beyond a certain limit (Hu et al. 2017, Moloisane et al. 2017). These issues might happen because of insufficient technologies used by organisational networks (Menelaou et al. 2017). An application of computerised mechanisms is needed to distribute traffic data to drivers in a timely way so that drivers can decide their routes

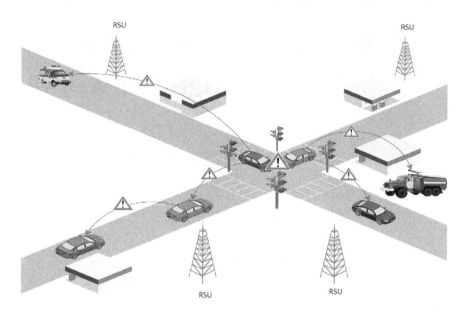

FIGURE 5.1 Vehicular ad-hoc network.

accordingly (Yan et al. 2017). This data includes roads with traffic jams, alternative roads, traffic congestion data, location, etc. (Bo et al. 2017). Hence, Intelligent Transportation System (ITS) was prominently created by using vehicular networks. Vehicular Ad-hoc Network (VANET) is a network of vehicles expressed as vehicle-to-roadside (V2I) and vehicle-to-vehicle (V2V) communication architecture to provide roadside services, navigation and safety information as shown in Figure 5.1.

Significant research in ITS enabled VANET towards an Internet of Vehicles (IoV), which assures innovative, commercial and technical capabilities (Talib et al. 2017). Recent development in the area of intelligence, computers, communications and control has led to tremendous growth in all intelligent devices which were equipped with embedded processors and wireless communication technologies. These smart devices were utilised to provide an approximate and safer platform by the Internet of Things (IoT). ITS is a smart traffic system that has the capacity to detect, notice and control traffic flows. It denotes the crucial factor concept in the smart city philosophy. An optimised traffic control system will minimise emission of gases and fuel consumption (Fernández-Ares et al. 2017). Such a system is based on vehicle mobility data and traffic data. Spread sensors like vehicular traffic, low sensor, GPS devices and smart meters will create serious technical and social effects. IoV is a global network for vehicles supported by Wireless Access Technologies (WAT). Its key concept is logistics, road infrastructure, vehicle safety and traffic control. IOV could be combined with technologies like Vehicular Cloud Computing (VCC), Network Function Virtualization (NFV), Software-Defined Networking (SDN), 5G and Fog computing to enhance the services (Fernández-Ares et al. 2017).

Cloud computing technology has emerged in several domains with the slogan pay as you use. The pools of resources are available virtually. Cloud computing in the transportation domain plays an important role in handling vehicles big data

(Talib et al. 2017). It allows people to think of autonomous driving, vehicle control and to design new intelligent systems. It is a promising technology for storing, communicating and computing vehicular data. Many researchers have proposed computation technologies to analyse the vehicular cloud network, urban traffic control, road safety measures, alerting passengers of accidents, object detection, cloud network security models and infotainment services, etc. (Bouyahia et al. 2017, Perova 2017, Menelaou 2017, and Talib et al. 2017). Vehicles at airport parking or at office buildings can offer to rent their vehicular cloud resources on demand. Hence the Internet can be used at any time, anywhere and anything to meet travellers' expectations. This leads to the exponential growth of data on clouds. Data in vehicular cloud networks can have 6V's of big data, listing: volume, velocity, variety, value, veracity and variability. On the basis of these 6V's vehicular data is validated and treated as big data (Jin et al. 2015). From the perspective of transportation planning, big data has various challenges (Milne and Watling 2019). The three main features of big data, huge-volume, multi-sources and fast-changing, make it difficult to use traditional methods for prediction. So there is a need for an algorithm which investigates its computability, computational complexity. Data is collected from various resources, sensors, roadside units, cameras, traffic signals, etc. (Zhu et al. 2018). Desired attributes may be missing from the collected data or may not be structured to be easily manipulated using data mining algorithms. Among the four stages of mining, data pre-processing is the first task to be computed. Classification algorithms are performed on the structured data to identify the classes of datasets for analysis (Assuncao et al. 2015). Naïve Bayes, Decision Tree and Random Forest are some classification algorithms. To derive the meaningful information from classified data, machine learning algorithms are suitable. Machine learning algorithms are able to extract information automatically without any human assistance (Anawar et al. 2018). Depending on the traffic patterns and requirements of decision-makers various machine learning algorithms are defined, supervised, unsupervised, semi-supervised and reinforcement algorithms (Bhavsar et al. 2019).

5.1.1 IMPACTS OF INTELLIGENT COMPUTING TECHNOLOGIES IN VANET

Safety, non-safety and infotainment applications are incorporated in today's vehicles using on-board units. These applications require bandwidth reservation strategies automatically for safety than non-safety and infotainment applications. Hence efficient and effective radio resource management strategies and intelligent algorithms are to be designed. In the developed countries infrastructure and mobility management are improved with time. However, in underprivileged countries' infrastructure, heterogeneity, data dissemination is still an issue and challenge. Imparting the new paradigms of communication and computing technologies can overcome these issues. In this section, we discuss the benefits of intelligent computing technologies in vehicular networks. Today's vehicular networks need to be intelligent, flexible and scalable to reduce traffic congestion on roads. VCC was introduced to obtain the Cloud Computing advantages in the transportation domain. It defines the on-board units (OBU) or roadside units (RSU) as cloud storage units. Vehicles with high computing power become cloud nodes and provide all its resources and services as a utility. Hussain et al. (2012) proposed three different vehicular cloud architectures: vehicular clouds (VC), vehicles using clouds (VuC) and hybrid clouds. VC is categorised as static clouds and dynamic clouds. Vehicles at the parking lot act as static clouds and rent out their resources on demand. Dynamic clouds are formed by moving vehicles in an ad hoc manner to share resources. In VuC, traditional clouds are used for services. Hybrid clouds are formed by merging both VuC and VC to enhance the service utilisation and to calculate traffic density and provide information in an optimal way (Sharma et al. 2015). Most of the cloud architecture research considers vehicle layer, communication layer and cloud interface layer's design.

Cloud computing is based on the concept of pay as you go. It provides services like Network-as-a-Service (NaaS), Storage-as-a-Service (StaaS), Infrastructure-as-a-Service (IaaS), Software-as-a-Service (SaaS), Platform-as-a-Service (PaaS), Containers-as-a-Service (CaaS), etc. These services can also be used for disaster management. Salim Bitam et al. (2015) proposed a cloud-supported gateway model, called gateway as a service (GaaS), which efficiently utilises the Internet usage experience for vehicular networks. Underutilised vehicle resources are aggregated in VC to form a cloud network. Jiafu Wan et al. (2014) proposed an integrated architecture of Vehicular Cyber-Physical Systems (VCPS) and Mobile Cloud Computing (MCC) known as VCPS and MCC Integration Architecture (VCMIA) to assist drivers. It defined different layers and their functionalities to improve Quality of Service (QoS) at the micro-level. Vehicular cloud networks do not cater to services that are delay-sensitive and have high mobility. Hence the delay-sensitive applications may be addressed with the introduction of edge computing. Raza et al. (2019) suggested merging fog computing with conventional cloud computing. It is a potential solution for several issues in current and future VANETs. They discuss merging fog with the cloud to facilitate low latency and high mobility with good connectivity and real-time applications. In addition, software-defined networks are integrated to enhance fog computing to improve flexibility, programmability and global knowledge of the network. They further provided two scenarios for the timely dissemination of safety messages in future VANET.

The high computing vehicle's resources are underutilised need to define algorithms so that computing resources can be given on a rent for parked vehicles. Some computing resources can act as relay networks to search nearby hospitals and police stations for emergency services. Vehicles as a Resource (VaaR) (Abdelhamid et al. 2015) is a proposed method to achieving it. VaaR provides storage and Internet facility as services. Hussain et al. (2012) defined VCC as three-layer architecture; in the first layer using inside sensors of car, driver's behaviour is collected. In the second layer, the collected data is sent to the cloud, manages the data, and provides different services. The communication layer is responsible for V2V communication using Dedicated Short Range Communication (DSRC), Wireless Access Vehicular Environment (WAVE) and cellular hybrid technology (Patil et al. 2018). Abdelhamid et al. (2015) proposed using vehicles parked at airport as data centres. The data generated by these data centres is needed to define the algorithm to generate structured data. Yassinea et al. (2019) present the current state of the research and future perspectives of fog computing in VANETs (Raw et al. 2013, 2015). The characteristics of fog computing and services provided for VANETs are discussed along with some opportunities, challenges and issues directing to potential future (Raw et al. 2012, El-Sayed et al. 2017). Some authors also presented real-time applications of VANET that can be implemented using fog computing (Singh et al. 2021a, b). Some applications use fog computing to improve the periodic communication between the nearby vehicles, vehicles and nearby roadside communication units (traffic and roadside lights). Heavy traffic generated with this communication requires huge storage, computation and infrastructure. Patil et al. (2020) have used both cloud and edge for these requirements. Further, fog computing is augmented to the cloud for delay-sensitive communications in high mobility vehicles. Yuan et al. (2018) proposed two-level edge computing architecture for automated vehicles to utilise the network intelligence at the edge fully. Base stations and autonomous vehicles are used as edge nodes for coordinated content delivery. They further have investigated the research challenges of wireless edge caching and vehicular content sharing. Simulated results significantly reduce the wireless bottlenecks, more beneficial for automated driving services.

5.1.2 Wireless Communication Technologies

The Federation of Communication Consortium (FCC) defined the first communication technology dedicated to transportation as DSRC (Li 2010). DSRC is 75 MHz of spectrum at 5.9 GHz. It supports vehicle-to-vehicle and vehicle-to-infrastructure communication using a variant of the IEEE 802.11a technology. DSRC would support safety-critical communications but fails to transfer the basic safety messages (BSM) when vehicle density increases, which has a high probability of colliding of the messages and leads to loss of reception of messages. WAVE (Amadeo et al. 2012) is a Wireless Access in Vehicular Environment used to improve packet delivery in congested vehicular ad-hoc networks. Later demand for vehicular entertainment and user applications leads to the use of hybrid communication technologies like cellular and Wi-Fi (802.11a/b/g/n and WiMAX) (Amadeo et al. 2012). DSRC and WAVE do not provide the required data rate and signal strength so it is necessary to switch to new technology that provides high data rate for long-distance connectivity. Hence cellular networks 2G, 3G and 4G-Long Term Evolution (LTE) and Let-A have emerged as new communication media for data dissemination in vehicular networks (Arora et al. 2019). The fifth-generation cellular network 5G has surpassed the previous generations of cellular networks in terms of prominence, delay tolerances and data transmission rate. LTE is a new paradigm shift in wireless technology with Evolved Packet Switched (EPS) system architecture. It is a registered trademark owned by European Telecommunications Standards Institute (ETSI). It supports both multicast and streams, with carrier bandwidth of 1.4–20 MHz (Arora et al. 2019). The drawback of LTE is handover time for calls is too long. Long Term Evolution-Advanced (LTE-A) is an improvement over LTE with a channel width up to 100 MHz. LTE-Advanced is a 4G cellular system with relay topology, 100 mbps data rate up to 350 kmph and minimum, of 1 Gbps data rate up to 10 kmph. It is a device-to-device communication technology, which is most required for vehicular networks (Chekkouri et al. 2018).

5.2 BIG DATA ANALYTICS

5.2.1 Data Mining Techniques in the VANET

This section of the chapter elaborates literature review on data mining. Several provocations of data mining are explained in Chen et al. (2014) and Najada and Mahgoub (2016). The key challenge faced during data mining is the structuring of the data. Data is divided into these categories: structured, unstructured and semi-structured. For an uncomplicated analysis of the data, a structured form of data is preferable. If the data is unstructured or semi-structured, it is converted to structured data using data filtering techniques (El Faouzi et al. 2011). The subsidiary challenge is data class imbalance wherein the data is classified under one of the assigned orders. Further, heterogeneous data collected from different sources is tedious to format (Yang et al. 2019). Traffic data would be collected from various sources like camera, roadside units, base stations and vehicles, and are need to address and bring into one format so that data are classified as intended. A significant rise in transport-related data due to rapid growth in rural-urban migration and globalisation implies exponential growth of traffic in big cities. Identifying suitable data mining techniques for collecting and analysing this data to yield useful perceptions is a challenging task. This section discusses different data mining techniques and identifies the suitable technique for vehicular big data. Data mining generates hidden data patterns in large datasets using scalable and efficient algorithms. Every data mining technique comprises seven steps which need to be followed: cleaning, integration, selection, transformation, data mining, knowledge presentation and pattern evolution (Jiawei et al. 2002). Big storage structures, like Hadoop, handle these big data, and can store large databases. However, data mining in large databases has innumerable requirements and challenges for researchers and developers. Dobre et al. (2014) classified different areas of data as web, text, and spatial, multimedia time series, biological, educational and ubiquitous data mining. Each type of data requires a different type of mining depending on the classified data's structure, behaviour and requirements. The following are some open-source tools for mining: WEKA, Rapid Miner, Orange, DataMelt, and Knime.

Data mining includes the following functionalities.

1. Classification: Classification is a decision-making method. Here unknown label data objects are identified and predicted by sets of models. Details are provided in Kesavaraj et al. (2013). Different classification methods include Bayesian network, Random Forest, decision tree induction, hierarchical classification, neural networks and support vector machines (De Sanctis et al. 2016, Karthick et al. 2019).
2. Clustering: Algorithms like K-means, SVD, CURB, etc. belong to this category. These algorithms divide data into groups of data that have the same meaning. Each group pattern is same and different in the other groups. Unsupervised learning is a part of clustering (Chen et al. 2014, Bedi et al. 2014).
3. Association: It is the set of rules, which displays frequently occurring data together in a given set of data as an attribute value (Han 2017, Javed et al. 2019).

4. Time series: Some of the applications are based on timestamp activity which is the main criteria for analysis and is considered as a primary attribute for classification. It comprises a set of methods and techniques for time series data (De Sanctis et al. 2016).

Hadoop Distributed File System (HDFS) is an open-source Linux-based framework. It permits the storage and processing of big data in a distributed environment across clusters of computers using programming models. It provides high-throughput access to applications and is mostly suited for big datasets (Chandra et al. 2016). HDFS has master/slave architecture. NameNode, a master server, manages file systems. Many other DataNode manages storage of files. Our research work uses HDFS for storing voluminous vehicle data. These data are programmed by using Map Reduce. It is a programming model with sets of key/vale pairs that are generated by mapping each logical record of the input. Applying a reduced operation to all the values that share the same key, the data is aggregated appropriately to derive the new information. Map and Reduce operations allow large computational operations to parallelise effortlessly and to use re-execution. Fault tolerance is one of the primary concerns of Map Reduce (Mneney et al. 2016).

Traffic data contains hidden values; meaningful information is extracted by applying data analytics algorithms; for example, vehicles' speeds can be collected to analyse congestion and the number of accidents from the roadside sensors. Analysis of vehicle waiting time at signals can be used to produce meaningful information to improve and optimise traffic light policies and flow. CCTV images will help in detecting and classifying objects and identifying their vehicle parts behaviours. A systematic review of big data in transportation is discussed in Jiawei et al. (2002).

The challenge of big data in transportation is data collection, data storage, data quality and security. Processing and operations can efficiently be performed by using algorithms. Big data in ITS has encouraged innovations of complex systems. Real-time modelling based on mobile trajectory data is proposed in Dobre et al. (2014) and a specific experiment is designed for analysis. Najada et al. (2016) generated different travel patterns for week days and analysed, using cellular networks using trip extraction. Initially, a small-scale cellular network dataset was collected along with GPS tracks from 20 mobile phones and analysed. Later they considered large cellular real-time datasets of Sweden city and analysed them.

In 2001 big data was defined with the three V's (Volume, Velocity and Variety), in 2013 its definition became data that does not fit into an Excel spreadsheet and later in 2016 two more V's were added, Variability and Veracity (Chen et al. 2014). In 2018 big data was defined with a sixth V (Value) using which knowledge can be extracted. Defining big data in transport planning perspectives has many challenges as defined in Chen et al. (2014). Not all challenges need to be addressed but they are part of planning. The volume of big data is complex to handle. To alleviate this issue upto some extent to enhance the computing ability of the systems. Multi-faceted data types (Variety), timely response requirements (Velocity) and variability in the data (Veracity) are the major challenges of big data. Due to the variety of data types, an application often needs to deal with traditional structured data, semi-structured or unstructured data. Scarcity of resources leads to delay in collection, storage and processing of big data for a time period. Lastly, it is challenging to identify the best data cleaning methods to remove the inherent unpredictability of data (Kesavaraj et al. 2013). Three big features of big data are huge-volume, multi-sources and frequent variability, making it difficult to use traditional methods for prediction. Hence, there is a requirement of an algorithm that investigates its computability, computational complexity and algorithms. Data collected from various sources need to be aggregated using data fusion algorithms.

Different data fusions, data association and decision fusion techniques are explained in De Sanctis et al. (2016). Data fusion is collecting data from various sources and combining these data to get meaningful information. Durrant-Whyte proposed fusion techniques for relations between input data sources; Dasarathy proposed fusion according to input/output, data types and nature. The most famous fusion algorithm widely used is Kalman filter. Data association can be performed by using K-Means and Probabilistic Data Association (PDA) techniques. A state estimation method gives the observations at a particular time. Different filtering techniques can be used. The Bayesian approach is the mostly used technique for filtering.

5.2.2 Machine Learning for VANET

Machine learning (ML) deals with the design and development of different types of learning algorithms. These algorithms will allow computers to learn from the data without giving clear instructions. ML algorithms can extract information without any manual help (Karthick et al. 2019). ML plays a vital role in ITS for learning data and providing useful information to make decisions. In ML, input data is divided into two sets, Training dataset and test dataset. The training dataset is used to learn hypothesis h, and test dataset is used to evaluate the learned hypothesis. Learning is based on the type of algorithm used (Jha 2015). Four different algorithms are defined in ML, supervised learning, unsupervised learning, semi-supervised learning and reinforcement learning.

Supervised learning in ITS: A set of algorithms, where training data includes required output. The system tries to learn from its input and output as a function, and then learns to predict the output. For illustration, considering a highway scenario, the input parameters obtainable are volume (i.e., number of vehicles per hour at particular time) and age of the driver; average traffic speed is to be calculated as output (Bhavsar et al. 2019). The learning algorithm utilises this information for automated training dataset to compute the speed from a given input. Classification and regression are two methods of supervised learning. The example of counting the number of vehicles and over speed of vehicles on the highway classification algorithm is used. Suppose estimating the average speed of the vehicles on the highway is based on past data, regression is used. Learning is possible in supervised learning if historical data is available (Chen et al. 2014, Bedi et al. 2014).

Unsupervised learning in ITS: Unsupervised learning identifies hidden patterns based on underlying unlabelled data. Clustering and association are two methods of unsupervised learning (Han et al. 2017). Clustering algorithms rely on mathematical models. It creates many clusters of the same data points. For example, to find an association between the age of a driver and blood alcohol level involved in a crash, when crash data of a highway are given, and also, to provide critical information of road checkpoints to reduce the crash time of the day. Association learning will play a vital role in these kinds of problems.

5.3 EDGE-ENABLED DATA GATHERING AND AGGREGATION

5.3.1 Data Gathering

This section elaborates possible functionalities handled by Edge Server (ES). ESs possess computing, communication and storage facilities at the end of the network. Three ESs are configured using EdgeCloudSim in this work. Vehicles register itself with its registration Id, speed, location and destination to its nearest ES. Since ES does not have enough capacity to hold vehicle details for a longer time, data is moved to the cloud for future use. ES can store the most recent registered vehicles details. Hence, timestamp of each vehicle is maintained. Figure 5.2 depicts the comparison of both cloud and edge computing technologies' performance in terms of average service time, average server utilisation time, average network delay and average processing time and thus proves that edge computing is most prominent technology for handling delay-sensitive applications. However, data gathered at ES needs to compute delay-sensitive information rather than computing data to cloud. This work's data generation and collection environment is vehicular data for 400 vehicles, with Vehicle Id, Speed, Location, Destination and Timestamp. These data need to be acquired using different techniques. Techniques used for data collection are log files, sensors, network data collection using crawlers, etc. The most widely used collection techniques are log files and CSV files. We used a log files method to collect data at each ES.

```
Edge Cloud Configuration..
Edge Cloud Configuration..
Scenario started at 14/05/2020 00:10:48
Scenario: TWO_TIER_WITH_EO - Policy: NEXT_FIT - #iteration: 1
Duration: 0 min (warm up period: 3.0 min) - #devices: 10
Creating tasks...Done,
Creating device locations...Done.
SimManager is starting...Done.
.........10.........20.........30.........40.........50.........60.....
....70.........80.........90.........100
# of tasks (Edge/Cloud/Mobile): 2249(0/2249/0)
# of failed tasks (Edge/Cloud/Mobile): 8(0/8/0)
# of completed tasks (Edge/Cloud/Mobile): 2241(0/2241/0)
# of uncompleted tasks (Edge/Cloud/Mobile): 0(0/0/0)
# of failed tasks due to vm capacity (Edge/Cloud/Mobile): 0(0/0/0)
# of failed tasks due to Mobility/Network(WLAN/MAN/WAN): 8/0(0/0/0)
percentage of failed tasks: 0.355714%
average service time: 0.851505 seconds. (on Edge: NaN, on Cloud: 0.851505, on Mobile: NaN)
average processing time: 0.175787 seconds. (on Edge: NaN, on Cloud: 0.175787, on Mobile: NaN)
average network delay: 0.675718 seconds. (LAN delay: NaN, MAN delay: NaN, WAN delay: 0.675718)
average server utilization Edge/Cloud/Mobile:
0.000000/0.060201/0.000000
average cost: 0.0$
Scenario finished at 14/05/2020 00:10:49. It took 1 Second
```

FIGURE 5.2 Comparison of intelligent computing technologies.

5.3.2 Data Aggregation

Data stored at edge nodes need to be utilised to generate different traffic patterns and intimate the same to drivers. To perform data analytics at edge nodes, data should be structured, non-redundant and aggregated. Hence aggregation of data is carried out by using HDFS. HDFS is an open-source distributed file system that is used to handle and manage big vehicular cloud data. Each execution of Hadoop creates different NameNode, DataNode, TaskTracker, JPS, JobTracker, etc. NameNode keeps a record of metadata of all DataNode. DataNode stores data in clusters in the file system. Map Reduce is a programming tool supported by HDFS to process large clusters of data. User requests are referred to as jobs in Map Reduce. When a job is executed, Map and Reduce tasks are created. The input data are divided into many tasks and processed in parallel and key-value pairs are generated by intermediate shuffling and later sorted by the framework. Vehicle's data are given as input to the Map-Reduce tool. A total of 816 input records are provided to Map module. After Mapping, 3,261 records are generated as Map output records. These many records are given to the Reduce model. Data redundancy is achieved by configuring Map Reduce with Term Frequency Inverse Document Frequency (TF-IDF) algorithm, because eliminating redundant information in the message is an important task or else more bandwidth is consumed unnecessarily. The Term Frequency Inverse Document Frequency algorithm works as below.

$$\mathrm{TF}(v) = (\text{number of times vehicle } v \text{ appears in document})/ \times(\text{Total number of vehicles } V \text{ in the document})$$

Where V = Total number of vehicles. Here $V = 400$, v = vehicle Id (KA01, KA02...)

IDF = $V*\log\_e$ (total number of vehicles/number of vehicles containing same vehicle Id (V))

Multiply both TF and IDF values. If Vehicle Id appearing in each document than IDF score is zero, then redundant Vehicle Id stored in HDFS are ignored. In Map Reduce, execution of aggregators is managed by TaskTrackers. A total of 3,342 tasks are aggregated. Once TaskTrackers receives an aggregator request, the JobTracker immediately initialises instances of aggregators and reduces the tasks using the information attached to the request.

5.4 EDGE-ENABLED SERVICE CONTENT PREFETCHING AND STORING

The objective of the chapter is to perform data analytics of vehicular networks using edge computing. In the previous section, data is gathered at ES and cloud server as well. Before computation ES has to prefetch all contents from cloud servers to reduce accessing time during computation. Hence, Edge cache is configured in EdgeCloudSim to fetch data from the cloud and hold it temporarily till computation, and then migrated to the cloud for efficient utilisation of edge storage. Figure 5.3 depicts the configuration of edge servers.

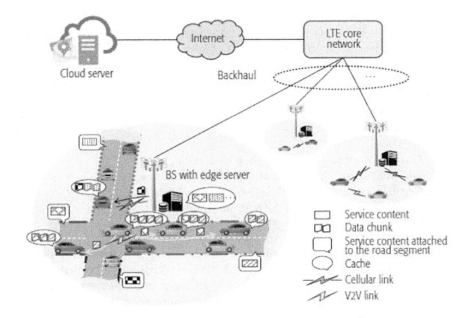

FIGURE 5.3 Configuration of edge server in EdgeCloudSim.

5.5 EDGE-ENABLED COMPUTING

This section focuses on data analytics in ESs. Data mining is a process of analysing the data from different perspectives to summarise useful information. Researchers are taking substantial efforts to solve this big data problem in the transportation system to help drivers and commuters. In Sections 5.1 and 5.2, data collection, storage and utilisation at ESs are discussed. Now HDFS file system contains structured and aggregated data, and it is necessary to extract meaningful information from these data. Hence, data mining is applied, and predictions are carried out to identify the most suitable ML algorithm for vehicular big data.

Classification of data is a process of figuring out what class a given observation belongs to one of the predefined classes. Naïve Bayes and Random Forest algorithms are used for classification in this work. Every classification algorithm has two phases, the learning phase and the prediction phase. In the learning phase the classifier trains its model for a given dataset, and in the prediction phase, it tests the classifier based on accuracy, error, precision and recall.

Naïve Bayes algorithm: Naïve Bayes is supervised learning algorithm; it predicts the class depending on the probability of belonging to that class. It uses the normal distribution to model numeric attributes. This algorithm is best suited for large datasets, and it is more accurate than other classification algorithms shown in the result section. It works on the probability factor. Basic formula for Naïve Bayes is:

$$P(A \mid B) + P(A \mid B) * \frac{P(A)}{P(B)}$$

where *A* is data tuple that belongs to a specified class and *B* is some hypothesis. For classification problems, we need to calculate $P(B|A)$, and probability of hypothesis is true if data instance is true. Similarly, $P(A|B)$ means probability of a given data instance is true if hypothesis is true.

$P(B)$ is probability of hypothesis *B* is true (independent of data) is known as prior independent of *B*.

$P(A)$ probability of data (independent of hypothesis) is known as prior independent. Steps to be followed for Naïve classifier are:

1. Calculate prior probability for given class instances.
2. Find similar probability with each attribute of the class.
3. Calculate posterior probabilities by using similar probability values.
4. Check which class has the highest probability; input will be belonging to that class.

Random forest algorithm: Random forest algorithm is a supervised learning algorithm. It builds an ensemble of decision trees, trained with the bagging method. The tree is built by searching for the best feature among a random subset of features. Hence, it generally results in a better model. In the result section, the accuracy of this algorithm is discussed.

5.6 RESULT AND DISCUSSION

Trained data is classified into five classes, namely, Major, Minor, Intermediate, No accident and ? (Class with erroneous data). Their prior probability is shown below. Prior probability is a probability calculated based on some hypothesis. This is the independent attribute.

```
Class Major prior probability = 0.25
Class Minor prior probability = 0.25
Class Intermediate prior probability = 0.25
Class No accident prior probability = 0.19
Class? prior probability = 0.06
```
The performance of classifier is measured by using confusion matrix.

```
=== Confusion Matrix ===
a b c d e <-- classified as
3 0 0 0 0  |  a = minor
0 3 0 0 0  |  b = major
0 0 2 0 0  |  c = no accident
0 0 0 2 0  |  d = intermediate
0 0 0 0 0  |  e =?
```

In the confusion matrix, three instances each belong to the major class and minor class; however, two instances each belong to no accident and intermediate class. Naïve Bayes does not have any error class instance. Table 5.1 shows prediction errors (if any) from both the classifiers. It is clear that Naïve Bayes has given 0 values for all classes.

TABLE 5.1
Summary of Naïve Bayes Classifier and Random Forest

Classifier Name	Correctly Classified Instances	Incorrectly Classified Instances	Kappa Statistic	Mean Absolute Error	Root Mean Squared Error	Relative Absolute Error	Root Relative Squared Error	Total Number of Instances	% of Classification
Naïve Bayes	10	0	1	0	0	0	0	10	100
Random Forest	784	6	0	0.0146	0.0891	88.7593	102.5195	790	99.2405

TABLE 5.2
Details of Accuracy

	Naïve Bayes Classifier				
Class	TP Rate	FP Rate	Precision	Recall	F-Measure
Minor	1	0	1	1	1
Intermediate	0	0	0	0	0
Major	1	0	1	1	1
No accident	1	0	1	1	1
?	0	0	0	0	0

TABLE 5.3
Details of Accuracy

	Random Forest Classifier				
Class	TP Rate	FP Rate	Precision	Recall	F-Measure
Active	0	0	0	0	0
Inactive	1	1	0.992	1	0.996

Where

- TP Rate: rate of True Positives, means instances are correctly labelled by the classifier.
- FP Rate: rate of False Positives, means instances are incorrectly labelled by the classifier.
- Precision: proportion of instances that are truly of a class divided by the total instances classified as that class.
- Recall: proportion of instances classified as a given class divided by the actual total in that class (equivalent to TP rate).
- F-Measure: A combined measure for precision and recall calculated as 2 * Precision * Recall / (Precision + Recall).

By comparing both the accuracy (Tables 5.2 and 5.3), Naïve Bayes classifier gives optimum solution and is also more accurate than the Random Forest classifier. Hence, Naïve Bayes classified dataset is given as input to the Logistic regression model to identify accidents and inform the ES automatically.

5.7 DATA ANALYSIS

Data analysis is a process of extracting useful information from the data and taking decision based on analysis. From the previous section, it is concluded that Naïve Bayes classification performs very well than random forest. The output dataset of the algorithm will be fed as input to the logistic algorithm. This method is known

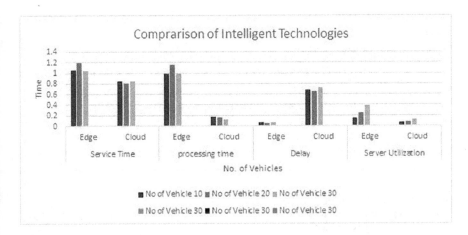

FIGURE 5.4 Number of alert messages delivered from simulation.

as the ensemble learning algorithm, where all learning algorithms will use the same dataset for prediction. Logistic regression is a supervised ML algorithm and is used when target is to be represented in categorical nature. For example, predicting if an accident happened or not. If the accident happened, value is 1 else 0. It is more efficient, updated easily for new data that is available and has less variance algorithm. Analyses of dataset are experimented by using Logistic regression by creating neural network layers. A number alert messages delivered to vehicles are shown Figure 5.4.

```
Below is output of simulation
Logistic regression
Scores of the neural networks node
Deep learning algorithm starting....
  Evaluation. ToSummaryString (title, false):
Summary
Correctly Classified Instances          36              100       %
Incorrectly Classified Instances         0                0       %
Kappa statistic                          1
Mean absolute error                      0.039
Root mean squared error                  0.0488
Relative absolute error                 12.9032 %
Root relative squared error             12.5888 %
Total Number of Instances               36

Evaluation. ToMatrixString ():
=== Confusion Matrix ===

 a b c d e    <-- classified as
 9 0 0 0 0  |  a = minor
 0 9 0 0 0  |  b = major
 0 0 9 0 0  |  c = intermediate
 0 0 0 0 0  |  d =?
 0 0 0 0 9  |  e = no accident
```

TABLE 5.4
Details of Accuracy

Class	TP Rate	FP Rate	Precision	Recall	F-Measure	ROC Area
Major	0.089	0.002	0.463	0.085	0.153	0.758
Minor	0.98	0.981	0.99	0.99	0.99	0.86
Intermediate	0	0	0	0	0	0.432
No accident	0	0.003	0	0	0	0.691
Weighted Avg.	0.985	0.942	0.971	0.976	0.9747	0.79

```
Optimising the node selection
NODE ID:19,27,35,54,90,104,127,163,195,200,227,244,268,293,299,
303,309,373,378,379
```

From Table 5.4, in no accident class, precision and recall value are zero; hence, F-measure is also zero. The accuracy of Logistic regression is 98%. This is the best prediction modelling technique for handling vehicular big data computed at the ESs.

5.8 CONCLUSION

Data generated by various vehicles on roads are underutilized. To extract meaningful information and utilize it for traffic controlling, providing alert messages to the vehicles automatically so that vehicles can take decisions accordingly. Hence, this chapter is focused on the integration of Edge computing with VANETs and performs data analytics. Edge computing technology has various benefits in vehicular networks in terms of efficient use of bandwidth, minimum latency and context awareness. The chapter discussed the survey on VANET, Cloud computing and Edge computing in the transformation domain. Three important concepts for edge computing are discussed. Firstly data gathering and utilisation at edge nodes in different formats and identifying most suitable format for storage on which different data mining techniques will classify the data. Secondly, data prefetching reduces data access time, stores data till analytics and disseminates messages since edge nodes have less data storage capability. Later the data will be moved to cloud servers for future reference. Lastly, use the data for analytics using ML algorithms. The simulation is carried out on Edgecloudsim and CloudSim simulation models. The results are proved that the efficiencies of the proposed system by using various measures such as delay and server utilisation are better than cloud networks. Comparing service time and processing time cloud networks is better. By improvising the edge computing capability still to minimise the delay and processing time by adding more processing capability power to edge will improve the system by merging with 5G.

Availability of data mining techniques and big data storage tools made an analysis of vehicle data meaningful to enhance road safety. We used Hadoop HDFS for storage of data, map reduce to aggregate and WEKA tool for mining to evaluate two classifiers on two big datasets. Naïve Bayes and Random Forest classifiers are used for classification; from experiments, Naïve Bayes gave the optimum results, with no errors and great accuracy compared to Random forest. Later logistic regression algorithm is applied along with a neural network to create a layer of networks to learn in deep to identify and classify the data. Results show logistic regression is the most suited algorithm. The work analysis will be a contribution and help to assist decision-makers in developing new rules.

REFERENCES

S. Abdelhamid, H. S. Hassanein, and G. Takahara. 2015. Vehicle as a resource (VaaR). *IEEE Network* 29: 0890–8044.

M. Amadeo, C. Campolo, and A. Molinaro. 2012. Enhancing IEEE 802.11 p/WAVE to provide infotainment applications in VANETs. *Ad Hoc Networks* 10: 253–269.

M. R. Anawar, S. Wang, M. Azam Zia, A. K. Jadoon, U. Akram, and S. Raza. 2018. Fog computing: An overview of big IoT data analytics. *Wireless Communications and Mobile Computing* 2018: 1–22.

A. Arora, A. Mehra, and K. Mishra.2019. Vehicle to vehicle (V2V) VANET based analysis on waiting time and performance in LTE network. *3rd International Conference on Trends in Electronics and Informatics (ICOEI)*, IEEE, Tirunelveli, 482–489.

M. D. Assunção, R. N. Calheiros, S. Bianchi, M. A. Netto, and R. Buyya. 2015. Big data computing and clouds: Trends and future directions. *Journal of Parallel and Distributed Computing* 79: 3–15.

F. Berzal, and N. Matín. 2002. Data mining: concepts and techniques by Jiawei Han and Micheline Kamber. *ACM SIGMOD Record* 31 (2): 66–68. Doi: 10.1145/565117.565130

P. Bhavsar, I. Safro, N. Bouaynaya, R. Polikar, and D. Dera. 2019. Machine learning in transportation data analytics. In M. Chowdhury, A. Apon and K. Dey (eds.), *Data Analytics for Intelligent Transportation Systems*. Elsevier, 283–307.

K. Bo, J. Teng, X. Liu, H. Liu, and H. Shi. 2017. Dissipating traffic congestion of emergency events through information guidance on mobile terminals. *Transportation Research Procedia* 25: 1276–1289.

P. Bedi, and V. Jindal. 2014. *Use of Big Data Technology in Vehicular Ad-hoc Networks*. 2014 International Conference on Advances in Computing, Communications and Informatics (ICACCI), IEEE, Delhi, 1677–1683, doi: 10.1109/ICACCI.2014.6968352

S. Bitam, A. Mellouk and S. Zeadally. 2015. VANET-cloud: a generic cloud computing model for vehicular Ad Hoc networks. *IEEE Wireless Communications* 22 (1): 96–102. doi: 10.1109/MWC.2015.7054724

Z. Bouyahia, H. Haddad, N. Jabeur, and S. Derrode. 2017. Real time traffic data smoothing from GPS sparse measures using fuzzy switching linear models. *Procedia Computer Science* 110: 143–150.

S. Chandra, and D. Motwani. An approach to enhance the performance of Hadoop map reduce framework for big data. *International Conference. on Micro-Electronics and Telecommunication Engineering*, 178–182. Doi: 10.1109/ICMETE.2016.64.

A. S. Chekkouri, A. Ezzouhairi, and S. Pierre.2018. A new integrated VANET-LTE-A architecture for enhanced mobility in small cells HetNet using dynamic gateway and traffic forwarding. *Computer Networks* 140: 15–27.

M. Chen, S. Mao, and Y. Liu. 2014. Big data: A survey. *Mobile Networks and Applications* 19 (2): 171–209.

J. Contreras-Castillo, S. Zeadally, and J. A. Guerrero-Ibanez. 2017. Internet of vehicles: Architecture, protocols, and security. *IEEE Internet of Things Journal* 5: 3701–3709.

J. Cui, H. Zhong, W. Luo, and J. Zhang. 2017. Area-based mobile multicast group key management scheme for secure mobile cooperative sensing. *Science China Information Sciences* 60: 098–104.

M. De Sanctis, I. Bisio, and G. Araniti. 2016. Data mining algorithms for communication networks control: Concepts, survey and guidelines. *IEEE Network* 30(1): 24–29.

X. Dobre. 2014. Intelligent services for Big Data science. *Future Generation Computer Systems*, 37, 267–281

N.-E. El Faouzi, H. Leung, and A. Kurian. 2011. Data fusion in intelligent transportation systems: Progress and challenges–A survey. *Information Fusion* 12: 4–10.

H. El-Sayed, S. Sankar, M. Prasad, D. Puthal, A. Gupta, M. Mohanty, et al. 2017. Edge of things: The big picture on the integration of edge, IoT and the cloud in a distributed computing environment. *IEEE Access* 6: 1706–1717.

A.J. Fernández-Ares, A. M. Mora, S. M. Odeh, P. García-Sánchez, and M. G. Arenas. 2017. Wireless Monitoring and tracking system for vehicles: A study case in an urban scenario. *Simulation Modelling Practice and Theory*. 73: ea22–42.

J. Grover, A. Jain, S. Singhal, and A. Yadav. 2018. Real-time VANET applications using fog computing. *Proceedings of First International Conference on Smart System, Innovations and Computing*, Smart Innovation, Systems and Technologies, Springer, Singapore, vol 79, 683–691. Doi: 10.1007/978-981-10-5828-8_65

J. Han. 2017. *Data Mining: Concepts and Techniques.* University of Illinois at Urbana-Champaign. Elsevier Publications.

A. J. Hasan, and A. Al-Omar. 2019. Traffic management system using VANET on cloud and smart phone. *2nd Smart Cities Symposium (SCS 2019)*, IET, Bahrain, 1–4. Doi: 10.1049/cp.2019.0219

Q. Hu, and L. Xu. 2017. Real-time road traffic awareness model based on optimal multi-channel self-organised time division multiple access algorithm. *Computers & Electrical Engineering* 58: 299–309.

R. Hussain, J. Son, H. Eun, S. Kim, and H. Oh. 2012. Rethinking vehicular communications: Merging VANET with cloud computing. *IEEE 4th International Conference on Cloud Computing Technology and Science.* 978-1-4673-4510-1.

M. A. Javed, S. Zeadally, and E. B. Hamida. 2019. Data analytics for cooperative intelligent transport systems, vehicular communications, *Elsevier Publications* 15:63–72.

V. Jha. 2015. Study of machine learning methods in intelligent transportation systems. UNLV Theses, Dissertations, Professional Papers, and Capstones, 2543. Doi: 10.34917/8220111

X. Jin, B. W. Wah, X. Cheng, and Y. Wang. 2015. Significance and challenges of big data research. *Big Data Research* 2: 59–64.

V. Karthick, and K.L. Sumathy. 2018. A similarity study of techniques in data mining and big data. *International Journal of Information and Computing* 6 (6): 174–181.

G. Kesavaraj and S. Sukumaran. 2013. A study on classification techniques in data mining. *The 4th International Conference on Computing, Communications and Networking Technologies (ICCCNT' 13)*, IEEE, Tiruchengode, 1–7.

Y. J. Li. 2010. An Overview of the DSRC/WAVE technology. In X. Zhang and D. Qiao (eds.), *Quality, Reliability, Security and Robustness in Heterogeneous Networks. QShine 2010. Lecture Notes of the Institute for Computer Sciences, Social Informatics and Telecommunications Engineering,* Berlin, Heidelberg, Springer, 74, 544–558. Doi: 10.1007/978-3-642-29222-4_38

J. Liu, J. Wan, D. Jia, B. Zeng, D. Li, C.-H. Hsu, et al. 2017. High-efficiency urban traffic management in context-aware computing and 5G communication. *IEEE Communications Magazine* 55: 34–40.

C. Menelaou, P. Kolios, S. Timotheou, C. G. Panayiotou, and M. Polycarpou. 2017. Controlling road congestion via a low-complexity route reservation approach. *Transportation Research Part C: Emerging Technologies* 81: 118–136.

D. Milne, and D. Watling. 2019. Big data and understanding change in the context of planning transport systems *Journal of Transport Geography* 76: 235–244.

J. Mneney, and J.-P. Van Belle. 2016. On traffic-aware partition and aggregation in map reduce for big data applications. *IEEE Transactions on Parallel and Distributed Systems.* 27: 818–828.

N.R Moloisane, R. Malekian, and D.C. Bogatinoska. 2017. Wireless machine-to-machine communication for intelligent transportation systems: Internet of vehicles and vehicle to grid. *40th International Convention on Information and Communication Technology Electronics and Microelectronics (MIPRO)*, Opatija, 411–415.

H. A. Najada, and I. Mahgoub. 2016. Big vehicular traffic data mining: Towards accident and congestion prevention. 2016 International Wireless Communications and Mobile Computing Conference (IWCMC), Paphos, 256–261. doi: 10.1109/IWCMC.2016.7577067

J. Patil, Suresha, and N. Sidnal. 2020. Edge Enabled Identification of Vehicle Emergency Situation using Data Aggregation and Filtering Approach. *International Journal of Advanced Science & Technology* 29: 4534–4543. http://sersc.org/journals/index.php/IJAST/article/view/24859

J. Patil, and N. Sidnal. 2018. *International Conference on Computing, Communication and Automation (ICCUBEA).* 978:5. IEEE Publication, Pune.

A. Perova. 2017. Methods of placement of business tourism centres in large cities as means providing traffic safety (on the example of St. Petersburg). *Transportation Research Procedia* 20: 487–492.

R. S. Raw, and S. Das. 2013. Performance analysis of P-GEDIR protocol for vehicular ad hoc network in urban traffic environments. *Wireless Personal Communication*, Springer, 68(1): 65–78.

R. S. Raw, S. Das, N. Singh, S. Kumar, and S. Kumar. 2012. Feasibility evaluation of VANET using Directional-Location Aided Routing (D-LAR) protocol. *International Journal of Computer Science Issues*, 9(5): 404–410.

R. S. Raw, D. K Lobiyal, S. Das, and S. Kumar. 2015. Analytical evaluation of directional-location aided routing protocol for VANETs. *Wireless Personal Communications*, Springer, 82(3): 1877–1891. Doi: 10.1007/s11277-015-2320-7.

S. Raza. S. Wang, M. Ahmed, et.al. 2019. *A Survey on Vehicular Edge Computing: Architecture, Applications, Technical Issues, and Future Directions.* Hindawi Publications, London. Article ID 3159762, 19 pages.

P. Singh, R. S. Raw and S. A. Khan. 2021a. Link risk degree aided routing protocol based on weight gradient for health monitoring applications in vehicular ad-hoc networks. *Journal of Ambient Intelligence Humanized Computing*, Doi: 10.1007/s12652-021-03264-z.

P. Singh, R. S. Raw, S. A. Khan, M. A. Mohammed, A. A. Aly and D.-N. Le. 2021b. W-GeoR: Weighted geographical routing for VANET's health monitoring applications in urban traffic networks. *IEEE Access*, Doi: 10.1109/ACCESS.2021.3092426.

M.S. Talib, B. Hussin, and A. Hassan. 2017. Converging VANET with vehicular cloud networks to reduce the traffic congestions: A review. *International Journal of Applied Engineering Research* 12: 10646–10654.

J. Wan, D. Zhang, Y. Sun, et al. 2014. VCMIA: a novel architecture for integrating vehicular cyber-physical systems and mobile cloud computing. *Mobile Network and Applications* 19: 153–160. doi: 10.1007/s11036-014-0499-6

G. Yan and D. B. Rawat. 2017. Vehicle-to-vehicle connectivity analysis for vehicular ad-hoc networks. *Ad Hoc Networks* 58: 25–35.

X. Yang, S. Luo, K. Gao, T. Qiao, and X. Chen. 2019. Application of data science technologies in intelligent prediction of traffic congestion. *Journal of Advanced Transportation* 2019: 1–14.

A. Yassine, S. Singh, M. S. Hossain, G. Muhammad. 2019. *IoT Big Data Analytics for Smart Homes with Fog and Cloud Computing.* Elsevier Publication.

Q. Yuan, H. Zhou, J. Li, Z. Liu, F. Yang, and X. (Sherman) Shen. 2018. Toward efficient content delivery for automated driving services: An edge computing solution. *IEEE Network* 32 (1): 0890–8044.

L. Zhu, F. R. Yu, Y. Wang, B. Ning, and T. Tang. 2018. Big data analytics in intelligent transportation systems: A survey. *IEEE Transactions on Intelligent Transportation Systems* 20: 383–398.

6 Impact of Various Parameters on Gauss Markov Mobility Model to Support QoS in MANET

Munsifa Firduas Khan and Indrani Das
Assam University

CONTENTS

6.1 Introduction 85
6.2 GM Mobility Model.. 86
6.3 Simulation Results 87
    6.3.1 Simulation Parameters... 87
    6.3.2 Experimental Results... 88
6.4 Results and Discussion 98
6.5 Conclusion and Future Work 99
References.. 100

6.1 INTRODUCTION

Mobile Ad-hoc Networks (MANETs) do not use any kind of infrastructure for communication. These provide huge application domains, including military areas, search and rescue operations and private fields [1]. The network provides a wide range of flexibility and reliability, making it advantageous and active for research. One of the major reasons for complex routing is node mobility which creates path failure. As a result, it becomes difficult to provide Quality of Service (QoS) in MANET [2–4]. QoS is the performance level of a service provided by a network to the user [5–7]. For the simulation of mobile nodes, it is necessary to use mobility models with any routing protocol [8]. One of the major reasons for choosing Gauss Markov (GM) mobility model for performance analysis with different parameters is that among so many existing mobility models presented by Meghanathan [9,10], Kumar et al. [11] and Guimarães et al. [12], it is concluded that GM mobility performs better in compared to other presented mobility models. However, Meghanathan [9] proposed an algorithm OptPathTrans to find stable paths between a particular source and destination node. Using mobility

DOI: 10.1201/9781003206453-6

models, namely GM and Random Waypoint on this algorithm, through simulation the connectivity of the network, route lifetime and hop count are analyzed. Simulation is done using discrete-event self-developed MANET simulator with a different number of nodes like 25, 75 and 100, varying node speed 5, 15 and 30 m/s and various values of tuning parameter α like 0, 0.2, 0.4, 0.6, 0.8 and 1. It is observed that the network connectivity is almost similar for both the mobility models using 75 and 100 nodes, but for 25 nodes the network connectivity is stronger using GM mobility model with the value of tuning parameter α as 0.4 and 0.8. The minimum hop count obtained using GM mobility model provides a route with less lifetime, whereas the network lifetime is better using Random Way point mobility model. It is also analyzed that with larger values of α stable routes can be obtained. Liang et al. [13] proposed a predictive distance-based mobility management scheme where the next position and velocity of a node are determined using a GM mobility model's probability density function based on the current node location and velocity. The proposed algorithm reduces the cost more than 50% for all the systems. Alenazi et al. [14] modified the existing 3D GM Mobility Model by adding a buffer zone to provide similar characteristics with the original 2D GM mobility model. The buffer zone is specified using parameters buffer thickness (Υ) and buffer ratio (β). The predefined values of Υ and β are 0 and 1, respectively. When a node reaches the buffer zone, it bounced back to the inner zone using mean direction and mean pitch values. Simulation is done in NS-3 using the value of tuning parameter α as 0, 0.25, 0.5, 0.75 and 1, the time-step is considered as 1, $\beta = 0.5$, 0.8 and 0.9 and distance random variance as 1 and speed random variance as 5. It is observed from the simulation result that node moves smoothly around the simulation boundary without bounces. The proposed model also works well in a 2D simulation area. The Ad-hoc On-Demand Distance Vector (AODV) routing protocol is used for the investigation of various simulation parameters with GM mobility model.

The organization of the chapter is as follows: Section 6.2 explains the GM mobility model. Section 6.3 shows the experiment and the results. Section 6.4 gives the results and discussions, and Section 6.5 is the conclusion.

6.2 GM MOBILITY MODEL

This mobility model is proposed by Liang and Haas [15]. "At a constant interval of time 't', the values of nodes with respect to speed, direction and pitch is calculated based on the former value of pitch, direction and speed at $(t-1)^{th}$ time interval" [8,9,16–19]. The following equations calculate the speed, direction and pitch values:

$$S_t = \alpha S_{(t-1)} + (1-\alpha)\tilde{S} + \sqrt{(1-\alpha^2)} Sx_{(t-1)} \qquad (6.1)$$

$$D_t = \alpha D_{(t-1)} + (1-\alpha)\tilde{D} + \sqrt{(1-\alpha^2)} Dx_{t-1} \qquad (6.2)$$

$$P_t = \alpha P_{t-1} + (1-\alpha)\tilde{P} + \sqrt{(1-\alpha^2)} Px_{t-1} \qquad (6.3)$$

"where S_t, D_t and P_t are the new speed, direction and pitch at time interval t, $\tilde{S}$, $\tilde{D}$ and $\tilde{P}$ are the mean speed, mean direction and mean pitch, Sx_{t-1}, Dx_{t-1} and Px_{t-1}, are random variables and α is a random variable whose value lie within the range of $0< \alpha <1$" [8,9,16–19].

When the value of α is minimal, it becomes memory less and equations (6.1–6.3) can be written as:

$$S_t = \tilde{S} + Sx_{t-1} \tag{6.4}$$

$$D_t = \tilde{D} + Dx_{t-1} \tag{6.5}$$

$$P_t = \tilde{P} + Px_{t-1} \tag{6.6}$$

Thus, the new speed, direction and pitch are determined using mean speed, mean direction and mean pitch and random variables Sx_{t-1}, Dx_{t-1} and Px_{t-1}, respectively [16].

When the value of α is maximal, it implies highest level of memory and equations (6.1–6.3) can be written as:

$$S_t = S_{t-1} \tag{6.7}$$

$$D_t = D_{t-1} \tag{6.8}$$

$$P_t = P_{t-1} \tag{6.9}$$

Thus, the new speed, direction and pitch are determined using the previous value of speed, direction and pitch at a time interval $(t-1)$ [16].

The next point of the mobile node is obtained using a new velocity vector where direction and pitch variables are considered as radians. The velocity vector is evaluated by the given mathematical equations [16]:

$$V_x = S_t \cos D_t \cos P_t \tag{6.10}$$

$$V_y = S_t \sin D_t \cos P_t \tag{6.11}$$

$$V_z = S_t \sin D_t \tag{6.12}$$

where V_x, V_y and V_z are the velocity vectors at X-, Y- and Z-coordinates of the node's position at time interval t. , S_t and P_t are the direction, speed and pitch of a node at time interval t.

6.3 SIMULATION RESULTS

6.3.1 SIMULATION PARAMETERS

AODV routing protocol is considered to check the impact of distinct parameters using GM mobility model. We have made huge experiment in NS-3 [20] on two groups of nodes, namely, smallest group and largest group, and the number of nodes considered for these two groups are 15 and 50, respectively. Many experiments have

TABLE 6.1
Simulation Environment

Parameters	Values
No. of nodes	15 and 50
Routing protocol	AODV
Number of flows	5
Total simulation time	120 seconds
Data rate	1,024 bps
Packet size	128 kbps
Propagation delay model	Constant Speed Propagation Delay
Propagation loss model	Friss Propagation Loss
Position allocator	Random Box X [0,170], Y [0,170], Z [0,150]
Mobility models	GM

TABLE 6.2
Parameters for GM Mobility Model

Parameters	Values
Bounds	X [0,170], Y [0,170] and Z [0,150]
Time-step	5, 10, 15 and 20 seconds
Tuning parameter α	0.25, 0.5, 0.75 and 1
Mean velocity	[0,20]
Mean direction	[0,6.283185307]
Mean pitch	[0.02, 0.5]
Normal velocity	Mean = 0.2, Variance=0.4 and Bound = 0.6
Normal direction	Mean = 0.5, Variance=0.6 and Bound = 0.7
Normal pitch	Mean = 0.3, Variance=0.5 and Bound = 0.8

been done with different values of parameters like time step and tuning parameters to investigate the performance of AODV using GM mobility model. Table 6.1 shows the simulation environment, and Table 6.2 represents the parameters and respective values for GM mobility model.

6.3.2 Experimental Results

The performance of the AODV is examined with the following QoS parameters throughput, Packet Delivery Ratio (PDR) and delay. We have simulated many times using different values of tuning parameter (α) as 0.25, 0.5, 0.75 and 1, time step as 5, 10, 15 and 20 seconds and two groups of nodes like 15 as the smallest group and 50 as the largest group to check the impact of these parameters on GM mobility model to investigate the QoS support in MANET.

- Throughput: We have considered four cases like A, B, C and D with different values of time step as 5, 10, 15 and 20 seconds, respectively. We have also taken different values of tuning parameters α like 0.25, 0.5, 0.75 and 1 for two sets of nodes 15 and 50. Different throughput values have been achieved for AODV using different parameters of GM mobility model as shown in Tables 6.3–6.6, respectively. All the cases are compared in Tables 6.7 and 6.8 and represented in Figures 6.1 and 6.2.

It is observed in Table 6.3 that for Case A, i.e. using time-step = 5 seconds, the highest throughput for AODV is 21.4966 kbps for 50 nodes using $\alpha = 0.5$, whereas the lowest throughput is 3.8484 kbps for 15 nodes using $\alpha = 0.25$. The throughput for AODV achieves better for 50 nodes in Case A.

TABLE 6.3
Throughput for Case A

Tuning Parameter	Number of Nodes	
	15	50
0.25	3.8484	13.5733
0.5	11.9608	21.4966
0.75	6.7795	19.0241
1	7.9763	16.5049

TABLE 6.4
Throughput for Case B

Tuning Parameter	Number of Nodes	
	15	50
0.25	15.7794	19.3668
0.5	4.8741	17.3625
0.75	15.4483	13.4597
1	5.6249	9.8559

TABLE 6.5
Throughput for Case C

Tuning Parameter	Number of Nodes	
	15	50
0.25	0.0689	15.5827
0.5	5.744	16.5349
0.75	7.4627	12.8523
1	19.9322	10.8508

TABLE 6.6
Throughput for Case D

Tuning Parameter	Number of Nodes	
	15	50
0.25	3.7705	10.7399
0.5	10.1769	13.9771
0.75	11.6717	9.7182
1	4.6426	12.7379

TABLE 6.7
Throughput for 15 Nodes

Tuning Parameter	Case A	Case B	Case C	Case D
0.25	3.8484	15.7794	0.0689	3.7705
0.5	11.9608	4.8741	5.744	10.1769
0.75	6.7795	15.4483	7.4627	11.6717
1	7.9763	5.6249	19.9322	4.6426

TABLE 6.8
Throughput for 50 Nodes

Tuning Parameter	Case A	Case B	Case C	Case D
0.25	13.5733	19.3668	15.5827	10.7399
0.5	21.4966	17.3625	16.5349	13.9771
0.75	19.0241	13.4597	12.8523	9.7182
1	16.5049	9.8559	10.8508	12.7379

It is observed in Table 6.4 that for Case B, i.e. using time-step = 10 seconds, the highest throughput for AODV is 19.3668 kbps for 50 nodes using $\alpha = 0.25$, whereas the lowest throughput is 4.8741 kbps for 15 nodes using $\alpha = 0.5$. The throughput for AODV achieves better for 50 nodes in Case B. However, it is also noticed that the throughput decreases with increasing value of the tuning parameter for 50 nodes.

It is observed in Table 6.5 that for Case C, i.e. using time-step = 15 seconds, the highest and lowest throughput for AODV is obtained for 15 nodes, which are 19.9322 kbps using $\alpha = 1$ and 0.0689 kbps using $\alpha = 0.25$, respectively. The throughput for AODV achieves better for 50 nodes in Case C. However, it is also noticed that the throughput is increasing with increasing value of tuning parameter for 15 nodes.

It is observed in Table 6.6 that for Case D, i.e. using time-step = 20 seconds, the highest throughput for AODV is 13.9771 kbps for 50 nodes using

Impact of Various Parameters on GM Mobility Model

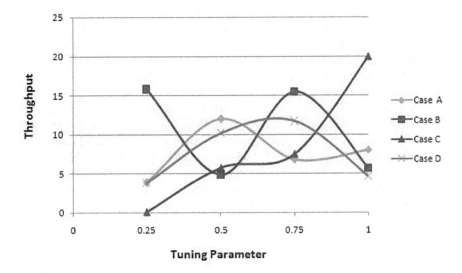

FIGURE 6.1 Throughput for a small set of nodes.

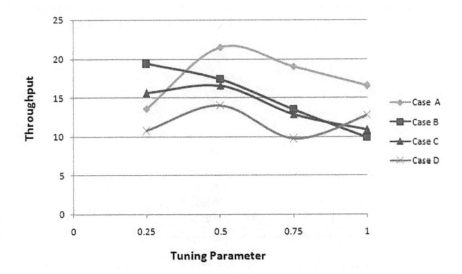

FIGURE 6.2 Throughput for a large set of nodes.

$\alpha = 0.5$, whereas the lowest throughput is 3.7705 kbps for 15 nodes using $\alpha = 0.25$. The throughput for AODV achieves better for 50 nodes in Case D.

For a small set of nodes, i.e. 15 nodes, the highest value of throughput for AODV is 19.9322 kbps using $\alpha = 1$, whereas the lowest is 0.0689 kbps using $\alpha = 0.25$ for Case C. It is noticed that for Case C, with the increasing value of tuning parameter, the throughput for AODV is also increasing, as shown in Figure 6.1. It is also observed that for the smallest set of nodes, the average highest throughput for AODV is achieved in Case B, whereas the average lowest throughput is obtained in Case D. Furthermore, it is also examined that the average throughput for AODV is higher using $\alpha = 0.75$ and the least average throughput achieved is using $\alpha = 0.25$ as shown in Table 6.7. On the other hand, for the largest set of nodes, i.e. 50 nodes, the highest value of throughput for AODV is 21.4966 kbps for Case A using $\alpha = 0.5$, whereas the lowest is 9.7182 kbps using $\alpha = 0.75$ for Case D as shown in Table 6.8. It is noticed in Figure 6.2 that for Case B and Case C with increasing value of tuning parameter, the throughput for AODV is decreasing. Additionally, it is evaluated that the average highest throughput for AODV is obtained in Case A, whereas the lowest throughput is obtained in Case D. Furthermore, it is examined that the average throughput for AODV is greater using $\alpha = 0.5$ and the average throughput is least using $\alpha = 1$. Moreover, it is also observed that the average throughput is better for a larger set of nodes for all cases than a small set of nodes.

- Delay: We have considered four cases like A, B, C and D for different values of time-steps as 5, 10, 15 and 20 seconds, respectively. We have also considered different values of tuning parameter α like 0.25, 0.5, 0.75 and 1 to obtain delay for AODV using GM mobility model with two sets of nodes like 15 and 50 as shown in Tables 6.9–6.12, respectively. All the cases are compared in Tables 6.13 and 6.14 and represented in Figures 6.3 and 6.4.

TABLE 6.9
Delay for Case A

Tuning Parameter	Number of Nodes	
	15	50
0.25	275.401	969.5180
0.5	445.6584	626.1533
0.75	250.7031	637.4992
1	264.9675	732.4455

TABLE 6.10
Delay for Case B

Tuning Parameter	Number of Nodes	
	15	50
0.25	225.8391	555.7945
0.5	279.3945	1,065.5534
0.75	215.7912	971.6491
1	308.5871	858.6993

Impact of Various Parameters on GM Mobility Model

TABLE 6.11
Delay for Case C

Tuning Parameter	Number of Nodes	
	15	50
0.25	357.6499	1,067.4834
0.5	277.8844	829.8965
0.75	177.5129	932.5075
1	316.4408	1,071.2125

TABLE 6.12
Delay for Case D

Tuning Parameter	Number of Nodes	
	15	50
0.25	248.0591	1,234.6738
0.5	301.7451	778.8636
0.75	353.7514	1,061.0042
1	300.6816	940.6426

TABLE 6.13
Delay for 15 Nodes

Tuning Parameter	Case A	Case B	Case C	Case D
0.25	275.401	225.8391	357.6499	248.0591
0.5	445.6584	279.3945	277.8844	301.7451
0.75	250.7031	215.7912	177.5129	353.7514
1	264.9675	308.5871	316.4408	300.6816

TABLE 6.14
Delay for 50 Nodes

Tuning Parameter	Case A	Case B	Case C	Case D
0.25	275.401	225.8391	357.6499	248.0591
0.5	445.6584	279.3945	277.8844	301.7451
0.75	250.7031	215.7912	177.5129	353.7514
1	264.9675	308.5871	316.4408	300.6816

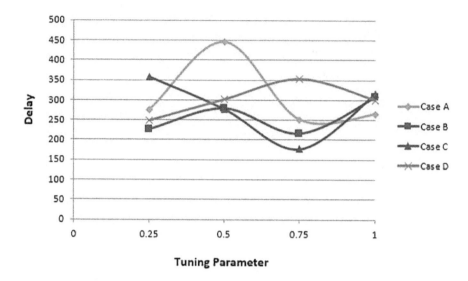

FIGURE 6.3 Delay for a small set of nodes.

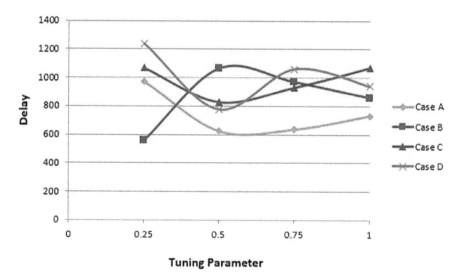

FIGURE 6.4 Delay for a large set of nodes.

It is observed in Table 6.9 that for Case A, i.e. using time-step = 5 seconds, the largest delay for AODV is 969.5180 seconds for 50 nodes using $\alpha = 0.25$, whereas the lowest delay is 250.7031 seconds for 15 nodes using $\alpha = 0.75$. The delay for AODV is achieved better for 15 nodes in Case A.

It is observed in Table 6.10 that for Case B, i.e. using time-step = 10 seconds, the largest delay for AODV is 1,065.5534 seconds for 50 nodes using $\alpha = 0.5$, whereas the lowest delay is 215.7912 seconds for 15 nodes using $\alpha = 0.75$. The delay for AODV is achieved better for 15 nodes in Case B.

Impact of Various Parameters on GM Mobility Model

It is observed in Table 6.11 that for Case C, i.e. using time-step = 15 seconds, the largest delay for AODV is 1071.2125 seconds for 50 nodes using $\alpha = 1$, whereas the lowest delay is 177.5129 seconds for 15 nodes using $\alpha = 0.75$. The delay for AODV is achieved better for 15 nodes in Case C.

It is observed in Table 6.12 that for Case D, i.e. using time-step = 20 seconds, the largest delay for AODV is 1,234.6738 seconds for 50 nodes using $\alpha = 0.25$, whereas the lowest delay is 248.0591 seconds for 15 nodes using $\alpha = 0.25$. The delay for AODV is achieved better for 15 nodes in Case D.

We have observed in Table 6.13 and Figure 6.3 that for a small set of nodes, i.e. 15 nodes, the minimum delay obtained for AODV is 177.5129 seconds for Case C using $\alpha = 0.75$, whereas the maximum delay obtained is 445.6584 seconds from Case A using $\alpha = 0.5$. Furthermore, it is also analyzed that the average minimum delay is obtained in Case B, whereas the average maximum delay is obtained for Case A. In addition to that, it is also observed that the average least delay is obtained using $\alpha = 0.75$ and the average highest delay is attained using $\alpha = 0.5$. On the other hand, for the largest set of nodes, i.e. 50 nodes, it is observed that the maximum delay for AODV is 1,234.6738 seconds for Case D using $\alpha = 0.5$, whereas the minimum delay for AODV is 555.7945 seconds for Case B using $\alpha = 0.25$ as shown in Table 6.14. It is also analyzed that the average minimum delay is achieved in Case A, whereas the average maximum delay is attained in Case D. In addition to that, it is also observed that the average least delay for AODV is obtained using $\alpha = 0.5$ and the average highest delay is attained using $\alpha = 0.25$. Moreover, it is also noticed that the average delay is better for a small set of nodes for all cases in comparing to the larger set of nodes.

- Packet-Delivery-Ratio: We have considered four cases like A, B, C and D for different values of time-steps as 5, 10, 15 and 20 seconds, respectively. We have considered two sets of nodes 15 and 50 for performance analysis using different values of α like 0.25, 0.5, 0.75 and 1. For each set of nodes, we have tested the GM mobility with different cases to achieve PDR for AODV as shown in as shown in Tables 6.15–6.18, respectively. All the cases are compared in Tables 6.19 and 6.20 and represented in Figures 6.5 and 6.6.

TABLE 6.15
PDR for Case A

Tuning Parameter	Number of Nodes	
	15	50
0.25	0.9233	0.8418
0.5	0.9526	0.8751
0.75	0.8781	0.8936
1	0.9176	0.8817

TABLE 6.16
PDR for Case B

Tuning Parameter	Number of Nodes	
	15	50
0.25	0.9492	0.8875
0.5	0.9263	0.8273
0.75	0.9131	0.8561
1	0.9720	0.8707

TABLE 6.17
PDR for Case C

Tuning Parameter	Number of Nodes	
	15	50
0.25	0.9623	0.8304
0.5	0.9523	0.8461
0.75	0.9064	0.8294
1	0.9309	0.8573

TABLE 6.18
PDR for Case D

Tuning Parameter	Number of Nodes	
	15	50
0.25	0.8394	0.8598
0.5	0.8500	0.8602
0.75	0.9007	0.8734
1	0.9470	0.8428

TABLE 6.19
PDR for 15 Nodes

Tuning Parameter	Case A	Case B	Case C	Case D
0.25	0.9233	0.9492	0.9623	0.8394
0.5	0.9526	0.9263	0.9523	0.8500
0.75	0.8781	0.9131	0.9064	0.9007
1	0.9176	0.9720	0.9309	0.9470

Impact of Various Parameters on GM Mobility Model

TABLE 6.20
PDR for 50 Nodes

Tuning Parameter	Case A	Case B	Case C	Case D
0.25	0.8418	0.8875	0.8304	0.8598
0.5	0.8751	0.8273	0.8461	0.8602
0.75	0.8936	0.8561	0.8294	0.8734
1	0.8817	0.8707	0.8573	0.8428

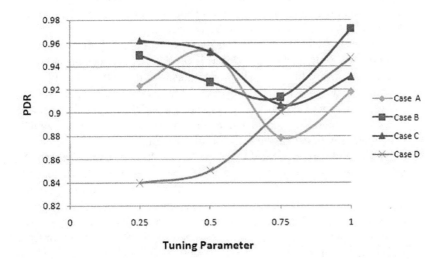

FIGURE 6.5 PDR for a small set of nodes.

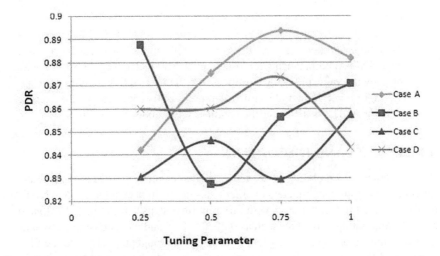

FIGURE 6.6 PDR for a large set of nodes.

It is observed in Table 6.15 that for Case A, i.e. using time-step =5 seconds, the highest PDR for AODV is 0.9526 for 15 nodes using $\alpha = 0.5$, whereas the lowest PDR is 0.8418 for 50 nodes using $\alpha = 0.25$. The PDR for AODV is achieved better for 15 nodes in Case A.

It is observed in Table 6.16 that for Case B, i.e. using time-step = 10 seconds, the highest PDR for AODV is 0.9720 for 15 nodes using $\alpha = 1$, whereas the lowest PDR is 0.8273 for 50 nodes using $\alpha = 0.5$. The PDR for AODV is achieved better for 15 nodes in Case B.

It is observed in the Table 6.17 that for Case C, i.e. using time-step =15 seconds, the highest PDR for AODV is 0.9623 for 15 nodes using $\alpha = 0.25$, whereas the lowest PDR is 0.8294 for 50 nodes using $\alpha = 0.75$. The PDR for AODV is achieved better for 15 nodes in Case C.

It is observed in Table 6.18 that for Case D, i.e. using time-step = 20 seconds, the highest PDR for AODV is 0.9470 for 15 nodes using $\alpha = 1$, whereas the lowest PDR is 0.8394 for 15 nodes using $\alpha = 0.25$. The PDR for AODV is achieved better for 50 nodes for lower value of tuning parameter, whereas for larger value of tuning parameter the PDR is better for 15 nodes in Case D. Moreover, it is also noticed that for 15 nodes, the PDR for AODV is increasing with increasing value of tuning parameter.

We have observed in the Table 6.19 that for 15 nodes, the highest PDR value for AODV achieved is 0.9720 for Case B using $\alpha = 1$, whereas the least PDR is 0.8394 for Case D using $\alpha = 0.25$. It is noticed in Figure 6.5 that with the growing value of tuning parameter, the PDR for AODV is rising for Case D. Moreover, it is observed that the average highest PDR is attained in Case B, whereas the average least PDR is achieved in Case D. It is also analyzed that the average highest PDR is obtained using $\alpha = 1$, whereas the average lowest PDR is obtained using $\alpha = 0.25$. On the other hand, it is noticed in Table 6.20 that the maximum PDR for AODV is 0.8936 for Case A using $\alpha = 0.75$, whereas the minimum PDR is 0.8273 for Case B using $\alpha = 0.5$. It is examined that the average PDR for AODV is higher in Case A, whereas average lower PDR is obtained in Case C. Moreover, it is also analyzed that $\alpha = 1$ gives the average highest PDR, whereas the average least PDR is obtained using $\alpha = 0.75$. It is also noticed in Figure 6.6 that with the growing value of tuning parameter the PDR for AODV is rising for Case A, but for $\alpha = 1$, it degrades. Furthermore, it is also noticed that the average PDR for AODV is better for the smallest set of nodes for all cases.

6.4 RESULTS AND DISCUSSION

We have taken GM mobility model for performance analysis with distinct values of time-steps and tuning parameters in order to check the impact of QoS in MANET. We have discussed the various cases with different time-steps values like 5, 10, 15 and 20 seconds, and tuning parameter such as 0.25, 0.5, 0.75 and 1 are considered to analyze the performance of AODV using QoS metrics like throughput, delay and the PDR. It is observed that for a small set of nodes, the average highest throughput is achieved in Case B where we have considered time-steps as 10 seconds, whereas the lowest

Impact of Various Parameters on GM Mobility Model

throughput is obtained in Case D where we have considered time-steps as 20 seconds. However, for large set of nodes, the average highest throughput is obtained in Case A where we have considered time-steps as 5 seconds, whereas the lowest throughput is obtained in Case D where we have considered time-steps as 20 seconds. It is observed that the maximum value of time-steps gives minimum throughput for both the set of nodes which implies that the lower value of the tuning parameter gives better QoS. Moreover, for smallest set of nodes, the average throughput is higher using the value of tuning parameter $\alpha = 0.75$ which is the second largest value of tuning parameter, whereas the least average throughput achieved is using tuning parameter $\alpha = 0.25$ which is the minimum value of tuning parameter which implies that the higher value of tuning parameter gives better QoS. On the other hand, for a larger set of nodes, the average throughput is greater using $\alpha = 0.5$, which is the average value of tuning parameter, and the least throughput is achieved using $\alpha = 1$, which is the maximum value of tuning parameter, i.e. QoS degrades with a maximum value of tuning parameter for a larger set of nodes. Additionally, it is also noticed that for a small set of nodes, the average highest PDR is attained in Case B where the value of time-steps that we have considered is average, whereas the average least PDR is achieved in Case D where the value of time-steps that we have considered is the maximum. However, for a larger set of nodes, the average PDR is better in Case A which takes the least value of time-steps and average lower PDR is obtained in Case C which takes the second larger value of time-steps, i.e. the higher value of time-steps leads to poor QoS for both the set of nodes. It is also analyzed that for small and large set of nodes, $\alpha = 1$ gives the average highest PDR which means the maximum value of tuning parameter gives the better QoS, whereas the average least PDR for large group of nodes is obtained using $\alpha = 0.25$, i.e. the lower value of tuning parameter degrades QoS and the average minimum PDR for the smallest set of nodes is achieved using $\alpha = 0.75$ which indicates the higher value of tuning parameter degrades QoS. Furthermore, it is also analyzed that for the smallest set of nodes, the average minimum delay is obtained in Case B where the value of time-steps that we have considered is average, whereas for large groups of nodes, the average minimum delay is better in Case A which takes the least value of time-steps. In addition to that, it is also observed that for the smallest set of nodes, the average least delay is obtained using $\alpha = 0.75$ and the average highest delay is attained $\alpha = 0.5$, whereas for a larger set of nodes, the average least delay is obtained using $\alpha = 0.5$ and the average highest delay is attained $\alpha = 0.25$.

6.5 CONCLUSION AND FUTURE WORK

In a MANET, nodes are mobile most of the time. Mobility models play an important role in MANET. It is used to determine a node position, speed and direction so that efficient communication can take place. We have taken GM mobility model for performance analyses considering four cases where different values of time-steps such as 5, 10, 15 and 20 seconds and tuning parameter such as 0.25, 0.5, 0.75 and 1 are considered to check the impact of QoS using QoS metrics like throughput, delay and PDR in routing protocol AODV. It is concluded that the distinct parameters give different results using the same mobility model. It is concluded that the maximum throughput, maximum PDR and minimum delay are obtained using time step 10 seconds for a small group of nodes, whereas using time step as 5 seconds for larger groups of nodes. Moreover, it is also noticed that for a small set of nodes, the tuning parameter $\alpha = 0.75$ gives the maximum throughput and minimum, whereas for a larger group of nodes, $\alpha = 0.5$ gives the higher throughput and minimum delay. However, the maximum value of tuning parameter, i.e. $\alpha = 1$, gives the highest PDR for both small and large set of nodes. This work will be helpful for researchers and student understand the impact of different parameters in MANET. It will make them understand how to select parameters for the mobility model for providing better QoS support and how the different parameters greatly impact QoS when the values of the parameters are changed using the same mobility model.

REFERENCES

1. Khan MF, and I Das. An investigation on existing protocols in MANET. In H. S. Saini et al. (eds.), *Innovations in Computer Science and Engineering, Lecture Notes in Networks and Systems* 74, pp. 215–224, 2019, © Springer Nature Singapore Pte Ltd, Singapore. Doi: 10.1007/978-981-13-7082-3_26
2. Raw RS, and S Das. "Performance analysis of P-GEDIR protocol for vehicular ad hoc network in urban traffic environments." *International Journal of Wireless Personal Communications*, Springer 68.1 (2013): 65–78.
3. S. Das, R. S. Raw and I Das. "Performance analysis of ad hoc routing protocols in city scenario for VANET." In *Proceedings of American Institute of Physics, (2nd International Conference on Methods and Models in Science and Technology*, 2011, pp. 257–261, DOI:10.1063/1.3669968.
4. Raw RS, DK Lobiyal, S Das, and S Kumar. "Analytical evaluation of directional-location aided routing protocol for VANETs." *Wireless Personal Communications*, Springer 82.3 (2015): 1877–1891, doi: 10.1007/s11277-015-2320-7.
5. Khan MF, and I Das. "Effect of different propagation models in routing protocols." *International Journal of Engineering and Advanced Technology (IJEAT)* 9.2 (2019): 3975–3980.
6. Khan MF, and I Das. "A study on quality-of-service routing protocols in mobile ad hoc networks." In *2017 International Conference on Computing and Communication Technologies for Smart Nation (IC3TSN)*. IEEE, pp 95–98, 2017.
7. Khan MF and D Indrani. "Implementation of random direction-3D mobility model to achieve better QoS support in MANET." *International Journal of Advanced Computer Science and Applications (IJACSA)* 11.10 (2020):195–203.
8. Camp T, J Boleng, and V Davies. "A survey of mobility models for ad hoc network research." *Wireless Communications and Mobile Computing* 2.5 (2002): 483–502.
9. Meghanathan N. "Impact of the Gauss-Markov mobility model on network connectivity, lifetime and hop count of routes for mobile ad hoc networks." *Journal of Networks* 5.5 (2010): 509.
10. Das S, and DK Lobiyal. "Effect of mobility models on the performance of LAR protocol for vehicular Ad hoc networks." *Wireless Personal Communications* 72.1 (2013, September): 35–48, Springer.
11. Kumar RS, and S. Pariselvam. "Formative impact of Gauss Markov mobility model on data availability in MANET‖." *Asian Journal of Information Technology* 11.3 (2012): 108–116.
12. Guimarães DA, EP Frigieri, and LJ Sakai. "Influence of node mobility, recharge, and path loss on the optimized lifetime of wireless rechargeable sensor networks." *Ad Hoc Networks* 97(2020): 102025.

13. Liang B, and ZJ Haas. "Predictive distance-based mobility management for multi-dimensional PCS networks." *IEEE/ACM Transactions On Networking* 11.5 (2003): 718–732.
14. Alenazi M, C Sahin, and JP Sterbenz. Design improvement and implementation of 3d Gauss-Markov mobility model. No. AFFTC-PA-12430. AIR FORCE TEST CENTER EDWARDS AFB CA, 2013.
15. Liang B, and ZJ Haas. "Predictive distance-based mobility management for PCS networks." In *IEEE INFOCOM'99. Conference on Computer Communications. Proceedings. Eighteenth Annual Joint Conference of the IEEE Computer and Communications Societies. The Future is Now (Cat. No. 99CH36320)*. Vol. 3. IEEE, New York, 1999.
16. Broyles D, A Jabbar, and PGJ Sterbenz "Design and analysis of a 3–D gauss-markov mobility model for highly-dynamic airborne networks." In *Proceedings of the International Telemetering Conference (ITC)*, San Diego, CA, 2010.
17. Biomo J-DMM, T Kunz, and M St-Hilaire. "An enhanced Gauss-Markov mobility model for simulations of unmanned aerial ad hoc networks." In *2014 7th IFIP Wireless and Mobile Networking Conference (WMNC)*. IEEE, Vilamoura, 2014.
18. Ghouti L, TR Sheltami, and KS Alutaibi. "Mobility prediction in mobile ad hoc networks using extreme learning machines." *Procedia Computer Science* 19(2013): 305–312.
19. Ariyakhajorn J, P Wannawilai, and C Sathitwiriyawong. "A comparative study of random waypoint and gauss-markov mobility models in the performance evaluation of manet." In *2006 International Symposium on Communications and Information Technologies*. IEEE, Bangkok, 2006.
20. Khan MF and D Indrani. "Performance evaluation of routing protocols in NS-2 and NS-3 simulators". *International Journal of Advanced Trends in Computer Science and Engineering (IJATCSE)* 9.4 (2020): 6509–6517.

7 Heterogeneous Ad-hoc Network Management
An Overview

Mehajabeen Fatima
SIRT, Bhopal

Afreen Khursheed
IIIT, Bhopal

CONTENTS

7.1 Introduction 103
    7.1.1 Wired and Wireless Communication Design Approach 105
    7.1.2 Enabling and Networking Technologies 106
    7.1.3 Taxonomy of HANET 106
7.2 Mobile Ad-hoc Network (MANET) 107
    7.2.1 Overview of MANET 108
    7.2.2 Simulation Results 109
7.3 Wireless Sensor Network (WSN) 112
    7.3.1 Overview of WSN 112
    7.3.2 Routing Protocol of WSN 115
7.4 Vehicular Ad-hoc Network (VANET) 117
    7.4.1 Characteristics 117
    7.4.2 Applications 118
7.5 Wireless Mesh Network (WMN) 118
7.6 Common Characteristics of HANET 119
7.7 Common Issues of HANET 120
7.8 Intelligent Management Requirement in HANET 120
References 121

7.1 INTRODUCTION

An ad-hoc network organization is an assembly of portable (mobile) nodes without any administrator or control point in which every node does versatile activity like router, handshaking, battery management etc. It is supposed that mobile nodes interact with each other in a dynamic network in which nodes can join and leave randomly. Presently complications are arising due to the structure of the present

Internet. Ad-hoc networks can contribute to address these difficulties. The two mobile nodes that are in coverage area of each other still have to use routers and switches to forward packets between them. Ad-hoc networks may be able to change this through direct connection of multiple wireless devices [1]. It is significant to note that the nodes should be compatible with present Internet too. For this reason, the standard hierarchical architecture, TCP/IP is implemented on all these nodes. Unlike present wired networks, ad-hoc networks have many additional restrictions like mobility, battery life, route discovery, maintenance and relatively lower processing [2].

The ad-hoc networks could be characterized into the first, second, and third eras. The original beginning of ad-hoc starts from 1972 called PRNET (Packet Radio Networks). CSMA (Carrier Sense Medium Access) and ALOHA (Areal Locations of Hazardous Atmospheres) were the methodologies utilized for medium access control. This is kind of distance-vector routing under PRNET. It was utilized on a preliminary premise to give diverse systems administration abilities in a battle environment. The IEEE 802.11 subcommittee had acknowledged the term ad-hoc networks. The examination of local area had begun to investigate the chance of sending data, especially in ad-hoc network in various areas. Then, the work was proceeded to improve the prior assembled ad-hoc network. GloMo (Global Mobile Information Systems) and the Near-term Digital Radio (NTDR) are improved forms and consequences of ceaseless endeavors. Global Mobile Information Systems were intended to furnish a climate with Ethernet type availability to interface anywhere and whenever to handheld devices [3]. Presently, ad-hoc network is viewed as the third era. Now in 2021, heterogeneous ad-hoc network (HANET) is looking like the demand of time. It is really required to develop it. The concept of HANET has applications in different extents like tactical networks, sensor networks, mobile networking, vehicular networking, emergency services, educational activities, online aids, etc.

Chapter organization is as follows:

First segment of the chapter consists of Wired and Wireless Communication Design Approach, Enabling Technologies, and Taxonomy. This part will give an introduction of the design technologies of networking. HANET's different technologies are Zigbee, Bluetooth, IEEE 802.11, and IEEE802.16. A comparison is presented in this section which will be based on data rate, range, configuration, and applications of these technologies.

The second segment of the chapter gives an outline of promising models of ad-hoc networks and discusses their distinguishing attributes. We specifically provide discussions on Mobile Ad-hoc Network (MANET), Wireless Sensor Network (WSN), Vehicular Ad-hoc Network (VANET), and Wireless Mesh Network (WMN). Different networks are taken individually in this chapter. An individual short discussion on different networks is provided below.

MANET does not have any fixed sender and receiver, no proper stations, no central incorporator and administrator, and no hardware structure. All nodes can leave and connect at any time and also connect in dynamic manner to each other. Thus, MANET is discussed in this chapter.

Nowadays, information on temperature, humidity, etc. is available in smart phones. Heart beat rate and blood pressure are made available through smart bands. All this happens due to accessibility of cheap cost sensors. The sensors could be connected with the support of a WSN. Therefore, WSN is taken in this chapter and a brief note is provided on this.

Ad-hoc network can be used to make vehicles smart and intelligent and this can be done with the help of VANET. VANET transforms each contributing vehicle into a router or node and allows vehicles to connect with each other in proximity of 300 m to 1 km. This will make a network with a wide reach. VANET empowers communication and connection among the roadside infrastructure and vehicles.

Since the vehicles commute are limited by roads and traffic guidelines. So, it is suggested to make a fixed structure at prominent areas to provide communication among vehicles. For this, a mesh type of network is required which can use the fixed infrastructure for communication. Thus, a WMN comes in picture.

WMNs are progressively independent and self-designed, with the nodes in the network self-building up an ad-hoc network and keeping up the lattice association. Mesh or Lattice networks utilize a combination of fixed and portable nodes interconnected by means of wireless connects to shape a multi-hop ad-hoc system.

The next part of this chapter highlights various common characteristics and research issues arising due to properties of ad-hoc network. The chapter closes with summary of the present requirements to manage data transmission and reception intelligently in HANET, e.g. improvement in data transmission, addressing the issues of MAC, routing, transport protocols, and security [4].

7.1.1 Wired and Wireless Communication Design Approach

The hierarchical layered approach was applied in the Open System Interconnect (OSI) reference model. It characterizes seven layers of stack. They are physical, data link, network, transport, session, presentation, and application layer. The seven layers are reduced to either five or four layers. This model is called TCP/IP. The accomplishment of the internet has led to the use of the layered architecture [1]. TCP/IP is a hierarchical model where convention stack layer works freely and trades data with nearby layers as it were. Correspondence between non-adjacent layers isn't permitted. The adjacent layers corresponding through static interfaces, independent of the individual network constraints and applications. The lower layer presents only a service interface to an adjacent layer and hides other information.

It was initially intended for wired connections, characterized by high data transmission, low latency, high Quality of service, static routing, less probability of packet loss, and no mobility [2]. In any case, in the wireless connections, execution and resources are restricted by insufficient accessibility of transmission range, limited coverage area, and employed modulation type etc. To resolve these issues, different MAC layer and network layer protocols and technologies are proposed. These technologies are discussed in next section.

7.1.2 Enabling and Networking Technologies

The single hop communication between devices is possible due to the presence of devices in direct range of each other. Multi-hop communication is used to connect the devices which are not in direct range. Different technologies are used for single and multi-hop communication. Heterogeneous ad-hoc networking uses the following communication technologies:

- Bluetooth (IEEE 802.15.1): It is used for individual area network.
- ZigBee (IEEE.15.4): It is featured by low data rate (<250 kbps) and short range (approx. 100 m).
- Ad-hoc Network (IEEE 802.11): High speed medium range MANETs.
- High speed wide range (IEEE 802.16).

IEEE 802.11 is most extensively used standard. There are several specifications in the 802.11 family. A comparison of 802.11 is given in Table 7.1.

Network utilizes the one hop direct communication capacities which are given by empowering innovations and reliable data transmission from sender to receiver. The sender and receiver may not be in direct coverage area. The sender and receiver direct-indirect coverage area communication is possible through different functions, provided by enabling technologies. These different functions are routing, channel sharing, channel handshaking, end-to-end reliable transmission of data, etc.

7.1.3 Taxonomy of HANET

HANET consists of single and multi-hop infrastructure-based and infrastructure less networks. Base station is used in infrastructure type of communication. Base stations

TABLE 7.1
IEEE Standard 802.11 Comparison

	IEEE 802.11a	IEEE 802.11b	IEEE 802.11g	IEEE 802.11n
Frequency band	5 GHz	2.4 GHz	2.4 GHz	2.4/5 GHz
Data rate	54 Mbps	11 Mbps	54 Mbps	600 Mbps
Bandwidth	20 MHz	20 MHz	20 MHz	20/40 MHz
Range	35 m	38 m	38 m	75 m
Modulation	OFDM	CCK modulated with QPSK	DSSS, CCK, OFDM	OFDM
Radio interference	Low	High	High	Low
Compatibility	Wi-Fi 5Mbps	Wi-Fi	Wi-Fi 11 Mbps	Wireless ethernet compatibility alliance
Mobility	Yes	Yes	Yes	Yes
Security	Medium	Medium	Medium	High

Taxonomy

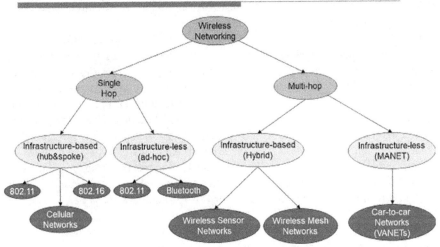

FIGURE 7.1 Wireless network taxonomy (https://images.app.goo.gl/u16cHq21kbxUd2h1).

or access points are fixed and centralized whereas infrastructure less networks (also called ad-hoc network) have no centralized access points. The nodes can move at anytime, anywhere from coverage area in ad-hoc network. Nowadays, dissimilar types of networking are available. So, a taxonomy of wireless networking is demonstrated in Figure 7.1.

The infrastructure-based wireless networks are sensor and mesh network. The infrastructure less networks are MANET and VANET. The following networks are mainly considered under HANET. Heterogeneous networks are discussed one by one in further sections.

1. MANET
2. VANET
3. WSN
4. WMN

7.2 MOBILE AD-HOC NETWORK (MANET)

The craving of peoples to be associated with each other at any place, at any time, and at any rate has prompted the improvement of wireless communication. This opens new panorama of exploration in ad-hoc networks. An ad-hoc network does not utilize any fix setup, any proper and fix station, and any fixed router. There is no brought together organization in ad-hoc networks. All nodes can connect and exit from coverage area whenever they want. All nodes can communicate with one another dynamically. All nodes can work as router, proficient to find routes and keep up routes, and can proliferate data packets in the network [4].

7.2.1 OVERVIEW OF MANET

MANET doesn't have any settled infrastructure. There is no single administrator. Every node works in conveyed distributed mode. Every node goes about as a free router and produces data information packets. MANET is self-configurable and does not depends on center management. The technologies allow mobile nodes to share data packets and attain present situation status [5]. Dedicated routers are not necessary. Every node forward packet of each other to facilitate information sharing between mobile hosts. Each node is free to connect and leave while communicating with other. Routes are formed and abolish in an unpredictable fashion. Consequently, the network status changes rapidly, the host loses the information of the present state [6]. Wireless channels are inclined towards error due to multipath fading, shadowing, interference initiating unusual connection transmission capacity, and latency. The distributed idea of MANETs implies that channel resources can't be given in a predetermined manner [7]. Ad-hoc network is demonstrated in Figure 7.2.

Different routing protocols are used for data transmission like proactive, reactive, and hybrid. Table 7.2 demonstrates main pros and cons of these protocols.

The routing protocols have different advantages and disadvantages. From Table 7.2 it is apparent that the reactive routing protocols are advance than others. Therefore, reactive routing protocols are analyzed. A comparative analysis of Ad-hoc On Demand Distance Vector (AODV) and Dynamic Source Routing (DSR) is done for mobility. Speed and pause time of a node defines the mobility. A comparison is given in section 7.2.2 for two protocols for different speeds keeping pause time constant and vice versa.

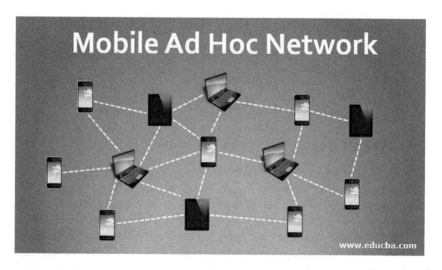

FIGURE 7.2 Ad-hoc mode.

TABLE 7.2
Pros and Cons of Different Routing Protocols

	Pros	Cons
Proactive [8]: e.g. Wireless Routing Protocol (1996, WRP), Destination Sequenced Distance-Vector (1994, DSDV), Cluster head Gateway Switch Routing (1997, CGSR), Optimized Link State Routing Protocol (OLSR)	Up-to-date routing information Route can be established quickly Small delay	Updates are propagated throughout the network routing information Constantly updated, this increases overhead [8] Loop creation tendency Large number of sources requirement Not scalable Insufficient use of routing information Large buffer size required
Reactive: e.g. Ad-hoc On Demand Distance Vector Routing (1999, AODV), Location Aided Routing (1998, LAR), Dynamic Source Routing (1996, DSR), Power-Aware Routing 1998 (PAR)	Route is searched on demand This saves resources This is Loop free Also eliminates need of continuous periodic table updates Destination sequence number is used to avoid stale information [8] Intermediate node utilizes the route cache This can be used for scalable network Required less buffer size Fast convergence	Not always have present routes status Large latencies Control traffic and overhead cost
Hybrid: e.g. Zone Routing Protocol (ZRP)	More scalability limited search cost Up-to-date routing information of routes	Arbitrary proactive scheme within zone Inter zone routing latencies More resources required for large size zone

7.2.2 SIMULATION RESULTS

Simulation results are demonstrated from Figures 7.3 to 7.8. The common scenario parameters are provided in Table 7.3.

A simulation is done for 60 node density and 15 links. Its impact is analyzed from Figures 7.3 to 7.5.

  i. Simulation is run for change in speed from 10 to 30 mps keeping pause time equal to 50 seconds.

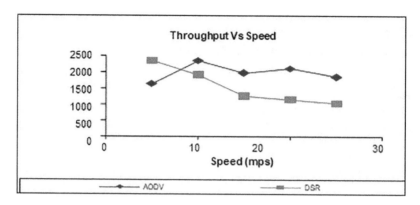

FIGURE 7.3 Throughput vs speed.

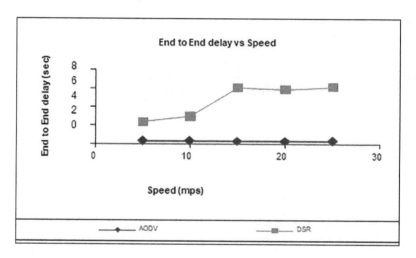

FIGURE 7.4 End to end delay vs speed.

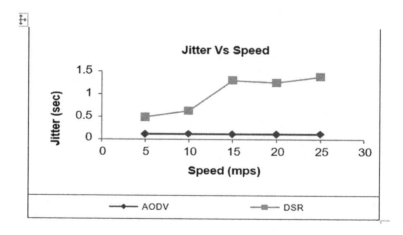

FIGURE 7.5 Jitter vs speed.

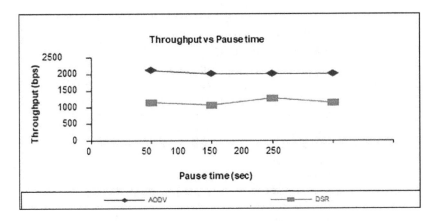

FIGURE 7.6 Throughput vs pause time.

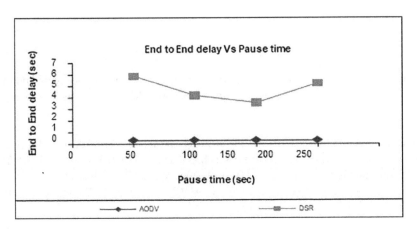

FIGURE 7.7 End to end delay vs pause time

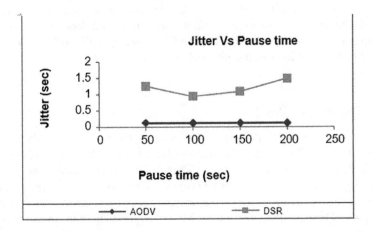

FIGURE 7.8 Jitter vs pause time

TABLE 7.3
Scenario Parameters

Simulation time	300 seconds
Number of nodes	60
Link	15
Simulation area	1,500 m × 1,500 m
Packet size	512 bytes
Channel	Path loss model
Energy model	Micaz
Battery model	Linear
Mobility model	Random way point
Data rate	2 Mbps
Node transmission range	300 m
Traffic model	CBR
Traffic	4,000

ii. Simulation is run for change in pause time from 50 to 200 seconds keeping speed equal to 20 mps. The links and nodes density are kept 15 and 60, respectively.

Figures 7.6 to 7.8 illustrate the high average throughput, low average end-to-end latency in AODV as compared to DSR. The AODV has jitter lesser than DSR. It is clear from the figures that the AODV performance is advance than DSR because of route search and maintenance procedure in AODV. Although AODV is better routing protocol, still many issues need to be resolved.

7.3 WIRELESS SENSOR NETWORK (WSN)

A WSN is a group of different types of sensors that can communicate wirelessly. Although the wireless sensors have limited resources like lack of memory, lack of computation power, less bandwidth, and low battery, still, with small physical size, it can be embedded in the physical environment and self-organizing ad doc networks [9].

7.3.1 Overview of WSN

Figure 7.9 illustrates WSN comprising of different sensors and their applications, but not limited to this. These sensor nodes, which are commonly termed as motes, primarily include numerous circulating, self-directed, minuscule, and ultra-low energy devices. Under the umbrella of this network, many distributed, battery operated devices also termed as nodes are embedded and further networked to collect as well as process and transfer data to operators. Moreover, it exhibits controlled capabilities of processing and computing.

Heterogeneous Ad-hoc Network Management

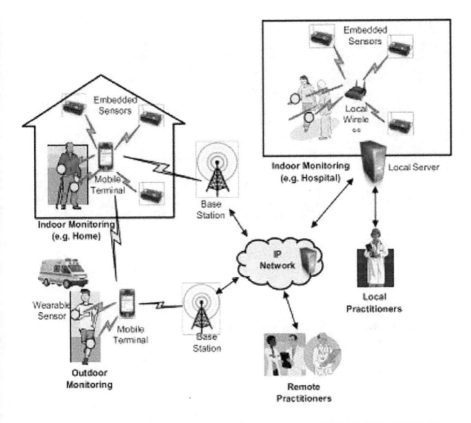

FIGURE 7.9 Wireless sensor network (https://images.app.goo.gl/rSunBE8XyGPXsj2K6).

A wide variety of sensors are available in the market. A wide range of system developments are possible with the help of sensors. Some types of sensors are given below. Pictures of some sensors, but not limited to this, are demonstrated in Figure 7.10.

- Pressure
- Temperature
- Humidity
- Light
- Biological
- Chemical
- Strain, fatigue
- Tilt
- Light
- Heart beat
- Color
- IR sensor
- Ultrasound sensor
- Real-time clock sensor, etc.

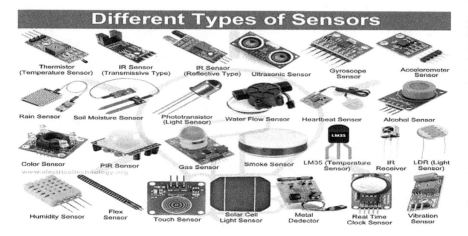

FIGURE 7.10 Different types of sensors (https://images.app.goo.gl/mUFAYPvLNk8WJPjE9).

Sensors have the following features:

- Able to endure harsh conditions (heat, humidity, stickiness, corrosion, erosion, contamination, and so on)
- No source of interference to frameworks being observed or potentially encompassing frameworks
- Could be deployed in huge numbers

Figure 7.11 demonstrate the applications implemented through sensors [9]. Potential applications for new intelligent systems are:

- Medical and healthcare
- Process monitoring and control
- Automation

FIGURE 7.11 Applications implemented with wireless sensor networks.

- Security and surveillance
- Environment monitoring
- Construction
- Smart homes

7.3.2 Routing Protocol of WSN

There are many routing protocols. A short brief of different routing protocols of sensor network is given below.

7.3.2.1 Sensor Protocol for Information via Negotiation (SPIN)

SPIN uses negotiation of resources and its adaptation. These **protocols** are designed to address flooding and gossiping. Negotiation reduces overlap and implosion and a threshold-based resource-aware operation is used to increase network lifetime [9].

7.3.2.2 Direct Diffusion (DD)

DD is a data-centric (DC) and application-aware protocol. The prime motive behind the adoption of DD is to coalesce the data entering the network from numerous sources by removing redundancy, lessening the transmissions count, thereby saving network energy, and enhancing lifetime.

7.3.2.3 Low Energy Adaptive Clustering Hierarchy (LEACH)

LEACH, the abbreviated form of Low Energy Adaptive Clustering Hierarchy, is primarily a cluster-oriented protocol based on the distributed cluster formation methodology. LEACH abruptly chooses the sensor nodes as cluster-heads and eventually rotates this responsibility so as to share and evenly distribute the energy load among various other sensor devices within the established network. Furthermore, in this protocol, cluster head CH compresses the data received from nodes belonging to the respective cluster and also sends an aggregated packet toward the base station. This strategy reduces the total amount of information which is to be transmitted to the base station [10].

7.3.2.4 Binary Scheme (BS)

BS follows a chain-based approach by classifying nodes into different levels. All the nodes, after receiving messages, increases level one toward the next level, consequently number of nodes is halved from one level to the next consecutive level.

7.3.2.5 Rumor Routing (RR)

RR protocol is a modified variant of DD protocol. It is explicitly used for application, not feasible for geographic routing. As a general rule DD makes use of flooding for injecting query to the network if there is no geographic criterion to diffuse tasks. In some cases when there is less data requested from nodes then the use of flooding is in vain. Thus, an alternative technique of flooding the events is taken up, provided that the event count is less than the query count. The key is to route the queries toward the node which shows a specific event rather than flood the entire network to retrieve the info for the event occurred.

7.3.2.6 Gradient-Based Routing (GBR)

GBR protocol implements energy balancing strategies as (i) using only the information from 1-hop neighbors at a time and (ii) an energy balancing approach that makes use of the cumulative energy of a full path. In both approaches, the energy link cost can be used for establishing a gradient of each sensor node [11].

7.3.2.7 Power Efficient Gathering in Sensor Information Systems (PEGASIS)

PEGASIS is a type of hierarchical routing protocol exploiting the greedy algorithm technique and chain-based approach. In this technique sensor nodes try to form a chain network. In case if a node dies in between, then that particular node is bypassed and the chain is reconstructed.

A comparative chart of different proposed protocols is given in Table 7.4.

Technique of routing in sensor networks is an emerging field with certain limitations but extremely promising results. Rumor routing has a trade-off between setup overhead and delivery reliability. GBR has potential for energy saving while realizing the network's routes, due to efficient back-off waiting scheme. LEACH and PEGASIS techniques show promising benefits such as scalability and efficient communication and maintain the energy consumption of sensor nodes. It performs data aggregation thereby reducing count of transmitted messages to base station [9].

TABLE 7.4
Comparison of Routing Protocols in WSN

Routing Protocols	Classification	Power Usage	Data Aggregation	Scalability	Query Based	Over Head	Data Delivery Model	QoS
SPIN	Flat/initiated/Data-centric	Ltd.	Yes	Ltd	Yes	Low	Event driven	No
DD	Flat/Data-centric/destination initiated	Ltd	Yes	Ltd	Yes	Low	Demand driven	No
RR	Flat	Low	Yes	Good	Yes	Low	Denand driven	No
GBR	Flat	Low	Yes	Ltd	Yes	Low	hybrid	No
LEACH	Hierarchicalt-initiated/Node-centric	High	Yes	Good	No	High	Cluster-head	No
PEGASIS	Hierarchical	Max	No	Good	No	Low	Chains based	No

Heterogeneous Ad-hoc Network Management

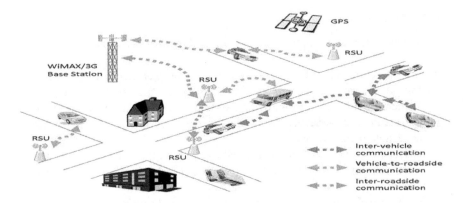

FIGURE 7.12 An example of VANET (https://images.app.goo.gl/HjugBmYY8WePVTbN6).

7.4 VEHICULAR AD-HOC NETWORK (VANET)

The VANET allows nodes to be joint or connected via wireless links; vehicles can freely ambulate in all directions. The salient feature is high mobility, dynamic topology, self-nodes organization, and interactions of vehicles [11–13]. Here in VANET because of larger node ambulatory and high vehicle velocity, the node spot location varies recurrently. An abrupt movement of nodes at a quick rate makes it difficult to judge the location of nodes and also raises concern for node privacy. VANET is ad-hoc in nature and assists the point nodes to collect data statics information from supplementary vehicles and street side units. An example of VANET is presented in Figure 7.12.

7.4.1 Characteristics

VANET is also abbreviated as VehAdhocNET. VANET has the following key features:

- Large Mobility: The vehicles in VANET are displacing quickly. This results in trivial method to foresee a node's location and creating assurance for node protection.
- Network Framework: Because of larger node ambulatory and arbitrary vehicle velocity, the node location varies recurrently.
- Wireless Communication: VANET is planned for the wireless surroundings. Nodes are concatenated and transfer data info to each other on a wireless platform.
- Boundless Network Dimension: VehAdhocNET can be practically realized for uno city, many cities, or for nations.
- Recurrent Swap over Data Information: The unplanned character of VANET inspires the nodes to accumulate data info from the vehicles and street units.

7.4.2 Applications

VANET has wide area of applications and some of them are given below:

- Infrastructure-Based Uses: effective enhancement of their management—transit; freeway management, intermodal stowage, crisis organization [11].
- Vehicle-Based Uses: for improving the street safety management, ride duration estimate, driver assistance during fog, crash prevention, collision avoidance, automatic settings.
- Driver-Based Services: in order to reduce the road traffic contention, street work data info, traveler's mode of payment. The hurdles in deployment is lack of availability of commonly available road side infrastructure, privacy of the vehicles, lack of coordination among manufacturing giants, Internet connectivity 24*7, and security [11,14–16].
- Travelers-Based Uses: for presenting novel administrations on-board-net access, distributed gaming, visitor data info, vacationer data, films declaration downloads.
- Driving experience and the safety of transportation [17].
- Safety Application (to enhance driving safety).
- EEBL—Emergency Electronic Brake Light (sudden braking).
- PCN—Post Crash Notification.
- RFN—Road Feature Notification (e.g. downhill curve).
- LCA—Lane Change Assistance.
- Commercial/Infotainment Application [17].
- Free flow Tolling.
- Social Networking.
- Finding Parking Spots.
- Dynamic Route/Travel Time Planning.
- Multimedia Content Exchange.
- Convenience Application (for better driving experience).
- Road Congestion Notification [17].

VANET applications are vast. So, this ad-hoc network is required to develop for making safe and easier life.

7.5 WIRELESS MESH NETWORK (WMN)

A WMN is a wireless setup consisting of multiple-hop communications in order to forward traffic for routing from one entry point to another as depicted in Figure 7.13. A mesh structure brings a hierarchy network in the architecture with the realization of devoted nodes (termed as routers-wireless) handshaking with one another and provided that wireless transport facilities from users to data traveling. Wireless-Mesh Net doesn't need central access points to direct the connection [18]. They are dynamically sole-planned and self-configured networks. Wireless Mesh multiple-hop network raises the effective area covered and link sturdiness of obtainable Wi-Fi's. WMNs nodes can be mobile or fixed. WMN may have Wireless Mesh routers, Mobile terminals or Stationary clients, Gateways, and Printers [19]. Basic types

Heterogeneous Ad-hoc Network Management

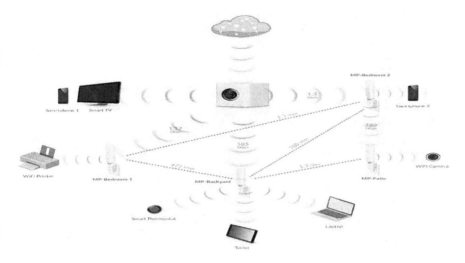

FIGURE 7.13 Mesh network.

of nodes are mesh-routers (MR) and mesh-clients (MC), where MR establishes an infrastructure backbone for clients. MCs are primarily cellular handheld-phones, portable laptops, and other wireless devices. MRs forward traffic to and from gateways without Internet-Connectivity.

MCs are supposed to be stagnant or ambulatory. It can generate a network for client mesh between themselves and together with MRs. Traffic is from user-toward-gateway, or user-toward-user.

7.6 COMMON CHARACTERISTICS OF HANET

- Self-Governing and Infrastructure Less: HANET doesn't rely upon any settled framework or concentrated organization. Every node works in shared mode, acts as a router, and creates autonomous information. A HANET's capacity to self-structure and self-oversee wipes out the requirement for central administration of organization [1]. HANET advances permit a power of versatile nodes to share the information more effectively and accomplish more prominent situational requirement.
- Mobility: Every node is independent to shift while handshaking with another nodes. Links can be created and shattered in a random way. Henceforth, the network grade changes swiftly, resulting hosts to have inaccurate knowledge of the existing network state.
- Multi-hop Routing: No committed routers are essential, each node acts as router and advances every others packets to enable information sharing among mobile hosts [2].
- Physical Traits: Wireless paths are intrinsically error-prone because of many-path fading, shadowing, intrusion, resulting in predictable link bandwidth and packet latency.
- Organizational Structure-Format: The dispersed nature of HANETs meaning that channel resources are not allocated in a programmed way.

7.7 COMMON ISSUES OF HANET

HANET is at the research stage and has many issues to resolve. The following areas are the main concerns in HANET regarding the design and performance.

A. Self-Governing: Absence of central administrative body to administer the function of the diverse moving nodes.
B. Self-Motivated Topology: Nodes are moving and can be coupled with dynamism in an uninformed way. Network Links fluctuate randomly and are governed by the proximity of one node to some other node.
C. Device Detection: To identify appropriate recently moved nodes and then to inform their existence, necessitate active update to smooth the progress of optimal path selection automatically.
D. Band-Width Trade-off Optimization: Wireless conduit have drastically subordinate capacity than the wired conduit. Lots of work is required to adjust bandwidth of channel.
E. Scarcity of Resources: Mobile nodes depend on battery power, which is a scant resource.
F. Scalability: Scalability is termed as the network ability to make available an up to standard level of service even in the existence of huge count of nodes in greater area with abundance of users. HANET is required to focus on it.
G. Inadequate Physical Security: Mobile nodes are subjected to greater security risks like peer2peer network design or a common wireless means easily reached to both legitimate network users and malicious attackers. Spoof, service attacks, and Eavesdropping should be considered.
H. Infrastructureless and Auto Healing: Auto healing trait demands HANET be supposed to rely onto manage any node touching out of its array.
I. Imperfect Class of Reception: Wireless communication has innate problem due to various error sources leading to dilapidation of the signals established. Ad-hoc addressing—Challenges in customary addressing technique need to be implemented.
J. Network Configuration: The entire HANET structure is self-motivated and is the cause for dynamic correlation and detachment of the changeable links.
K. Topology Safeguarding: Information upgradation of dynamic links between nodes in HANETs is a chief dare.

7.8 INTELLIGENT MANAGEMENT REQUIREMENT IN HANET

In HANET, a few troublesome issues emerge because of public-sharing character of the non-wired means, restricted broadcast power, restricted transmission broadcast range, mobility of node, and battery constraints, data transfer capacity limit, and so on. The restricted transmission scope of wireless network combined with the exceptionally powerful directing framework needs additional consideration. Mobility has lots of concern. It is needed to deal with the mobility, coverage, and battery power. The issues like dynamic routing, proficient access to channel and QoS maintain, bandwidth, harmonization, disseminated character, not have coordination centralized are required to take into consideration [20–22]. In wireless ad-hoc networks, the aforementioned issues are taken into consideration by Network layer and MAC Layer. Different protocols are used for MAC layer and Network layer to establish the path for data broadcast. The transport layer manages end-to-end delivery. Also, possible energy and power management is required as all nodes are battery-dependent.

The HANET is trying to manage through TCP/IP architecture. TCP/IP uses five layers for communication. The physical layer along with the MAC layer and routing layer collectively compete for the reserves of network in a wireless network. The data rate and transmission energy at physical layer are influencing the MAC layer activities and routing decisions. Scheduling and allocation of the non-wired channel is the responsibility of MAC layer. It creates the accessible transmitter bandwidth and the packet delay. Routing layer interruption to decide on the link also establishes concern. The destination of data packets sent to route is decided by routing layer. The contention level will be changed by routing decision at the MAC layer and consequently the parameter of physical layer. A stringent layered design is trying to deal with the next-generation communications dynamics which will be dominated by heterogeneous ad-hoc network. To make a move for standardization of heterogeneous ad-hoc network requires establishment of novel models whose potential applications have to be explored. Various challenges draw the attention of researchers toward the area where security and economics are required to ensure cooperation among nodes.

The aforementioned novel models endow with various pioneering application, and on the other hand their richness is yet to be explored. Quite a few added challenges have need of consideration from researchers. Novel models for security and economics needed to make certain of collaboration among nodes. Ad-hoc cloud needs novel strategies for ensuring provision of service, market-based resource management, data privacy, interoperability, etc. [23]. Researcher's hard work requires for the improvement in transmission of data and issues related to addressing in MAC, routing. For HANET, the research is at preliminary level and hard work is obligatory to develop methodology to address issues concerning management of spectrum, signaling, mobility management, security issues, and transport protocol. To conclude, novel models of application, tools related implementation, analytical models and benchmarks need development to evaluate the guarantee of upcoming models of ad-hoc networks.

REFERENCES

1. R. Ramanathan, and J. Redi. "A brief overview of ad hoc networks: challenges and directions", *IEEE Communications Magzine* Vol. 40, pp. 611–617, 2002.
2. D. Durich, and D. Montesinos. "E-book on adhoc networks telecommunication systems & networks", www.Comtel.Cz/Files/Download.Php.Id=5519.
3. J. Jubin, and J. D. Tornow. "The DARPA packet radio network protocols", *Proceedings of IEEE*, Vol. 75, pp. 21–32, 1987.
4. E. M. Belding-Royer, and C. E. Perkins. "Evolution and future directions of the ad hoc on-demand distance-vector routing protocol", *Journal of Ad Hoc Networks Elsevier Publication*, Vol. 1, pp. 125–150, 2003.

5. I. Chlamtac, M. Conti, and J. J.-N. Liu. "Mobile ad hoc networking: imperatives and challenges", *Journal of Ad Hoc Networks Elsevier Publication*, Vol. 1, pp. 13–64, 2003.
6. F. Dressler. "A study of self-organization mechanisms in ad hoc and sensor networks", *Journal of Computer Communications, Elsevier Publication*, Vol. 31, pp. 3018–3029, 2008.
7. A. Dollas, I. Ermis, I. Koidis, I. Zisis, and C. Kachris. "An open TCP/IP core for reconfigurable logic", In *Proceedings of the 13th Annual IEEE Symposium on Field-Programmable Custom Computing Machines (FCCM'05)*, IEEE, Napa, CA, pp. 297–298, 2005. doi: 10.1109/FCCM.2005.20
8. H. Badis, and A. Rachedi, Modeling tools to evaluate the performance of wireless multi-hop networks. In *Modeling and Simulation of Computer Networks and Systems*, pp. 653–682, 2015.
9. M. S. BenSaleh, R. Saida, Y. H. Kacem, and M. Abid, "Wireless sensor network design methodologies: a survey", *Journal of Sensors*, vol. 2020, Article ID 9592836, 13 p, 2020.
10. N. N. Malik, W. Alosaimi, M. I. Uddin, B. Alouffi, and H. Alyami, "Wireless sensor network applications in healthcare and precision agriculture", *Journal of Healthcare Engineering*, vol. 2020, 9 p, 2020.
11. R. Ghori, K. Z. Zamli, N. Quosthoni, M. Hisyam, and M. Montaser, "Vehicular ad-hoc network (VANET): review", In *IEEE International Conference on Innovative Research and Development (ICIRD)*, Bangkok, pp. 1–6, 2018.
12. S. Das, R. S. Raw, and I. Das, "Performance analysis of ad hoc routing protocols in city scenario for VANET", In *Proceeding of American Institute of Physics, (2nd International Conference on Methods and Models in Science and Technology: ICM2ST-11*, pp. 257–261, 2011, DOI:10.1063/1.3669968.
13. S. Das, R. S. Raw, I. Das, and R. Sarkar, "Performance analysis of routing protocols for VANETs with real vehicular traces", In D. P Mohapatra et al., (eds) *Advances in Intelligent Systems and Computing*, Vol. 243, pp. 45–56, 2014, DOI: 10.1007/978-81-322-1665-0_5.
14. R. S. Raw, D. K Lobiyal, and S. Das, "A probabilistic analysis of border node based MFR routing protocol for vehicular ad-hoc networks", *International Journal of Computer Applications in Technology, Inderscience*, Vol. 51, No. 2, pp. 87–96, 2015.
15. R. S. Raw, D. K. Lobiyal, S. Das, and S. Kumar, "Analytical evaluation of directional-location aided routing protocol for VANETs", *Wireless Personal Communications*, Springer, Vol. 82, No. 3, pp 1877–1891, 2015.
16. R. S. Raw, and S. Das, "Performance analysis of P-GEDIR protocol for vehicular ad hoc network in urban traffic environments", *Wireless Personal Communication*, Springer, Vol. 68, No. 1, pp. 65–78, 2013.
17. J. Jeong, Y. Shen, T. Oh, S. Céspedes, N. Benamar, M. Wetterwald, and J. Härri, "A comprehensive survey on vehicular networks for smart roads: a focus on IP-based approaches", *Vehicular Communications*, Vol. 29, ISSN 2214-2096, 2021.
18. J. Wang, B. Xie, and D. P. Agrawal. "Journey from mobile ad hoc networks to wireless mesh networks", In: Misra S., Misra S. C., Woungang I. (eds) *Guide to Wireless Mesh Networks. Computer Communications and Networks*. Springer, London, 2009.
19. M. O. Kabaou, and H. Hamouda. "Implementation and evaluation of opportunistic routing protocols for wireless and new generation communication networks", *Wireless Personal Communications* 112, 1165– 1183, 2020.
20. O. Kaiwartya, R. S. Raw, A. H. Abdullah, and Y. Cao, "T-MQM: testbed based multi-metric quality measurement of sensor deployment for precision agriculture-a use case", *IEEE Sensors Journal*, Vol. 16, No. 23, 8649–8664, 2016.

21. A. Ahmed, A. Abdul Hanan, K. Omprakash, D. K. Lobiyal, and R. S. Raw, "Cloud computing in VANETs: architecture, taxonomy and challenges", *IETE Technical Review*, Taylor & Francis, Vol. 35, No. 5, 1–25, 2017.
22. K. Rana, S. Tripathi, and R. S. Raw, "Inter-vehicle distance-based location aware multi-hop routing in vehicular ad-hoc network", *Journal of Ambient Intelligence and Humanized Computing*, Springer, Vol. 11, No. 11, pp. 1–13, 2020.
23. A.K. Yadav, R. Bharti, and R.S. Raw, "*SA^2-MCD*: secured architecture for allocation of virtual machine in multitenant cloud databases". *Big Data Research: An International Journal*, Elsevier, Vol. 24, 2021 Doi: 10.1016/j.bdr.2021.100187.

8 Deployment of the Biometrics-as-a-Service (BaaS) Design for the Internet of Biometric Things (IoBT) on the AWS Cloud

Vinayak Ashok Bharadi, Trupti S. Kedar
Mumbai University, FAMT

Pravin S. Jangid
LRTCOE

Mamta Meena
Vikrant Institute of Technology and Management

CONTENTS

8.1	Introduction	126
8.2	Strengthening Security of Transactions through Blockchain DB	126
8.3	Biometric Software as a Service (BAAS)	127
8.4	BAAS and Cloud Biometrics	127
8.5	Existing Work	127
	8.5.1 Biometric Trait Capture and Preprocessing	128
	8.5.2 Extraction of FVs	128
	8.5.3 Matching	128
	8.5.4 Decision	128
	8.5.5 Classification	128
8.6	Modification of Existing System	129
8.7	BAAS Deployment on Amazon AWS Cloud	129
8.8	IoBT Backend	131
	8.8.1 Step 1: Login to Your AWS Console and Create Instance	131
	8.8.2 Step 2: Login, Configure and Run	134
	8.8.3 Step 3: Build and Run Models on AWS	134

DOI: 10.1201/9781003206453-8

8.8.4 Step 4: Close Your EC2 Instance.. 134
8.9 Proposed System and Initial Deployment Results.......................... 135
8.10 Conclusion 137
Acknowledgments... 137
References... 138

8.1 INTRODUCTION

Biometric implies recognizing natural body structure identified with an individual or conduct highlights. Biometric system identifies human by consistent localization and feature extraction of a body part or the behavioral aspect of the person. A biometric method is a framework, which perceives the example that recognizes the individual people by their explicit body-structure-related aspects or behavioral feature which are quantifiable [1–3].

Various biometric traits are now available for human identification. Some examples include iris, face, fingerprints, palm prints, finger knuckle prints, voice, and signature. These traits are classified as physiological and behavioral [3]. Further the biometric system has two operational modes, recognition mode for 1:N matching and verification mode for 1:1 mapping [4]. Biometric systems have evolved over the decade, with the advancement in the smartphone, tablet and other devices with biometric sensors onboard. These devices are accessible to all, and with the proliferation of cloud-based application development, the storage and processing of the biometric traits over the cloud is the next widely adopted approach [4]. Infrastructure as a Service is the preferred cloud development model by the enterprise application developers, and it serves as foundation for the Platform as a Service model. The end application follows Software as a Service (SaaS) model. SaaS model is highly scalable and suitable for the biometric systems working at large scale [4]. Internet of Things (IoT) has various entities such as people, devices and their intercommunication for a particular set of services. When this interaction is happening for the biometric authentication of specific device or person over the Internet, it is termed as Internet of Biometric Things (IoBT). When biometric authentication for the smart applications is deployed on a cloud-based SaaS software model, it is referred as IoBT with Biometric as a Service (BaaS) at the backend.

8.2 STRENGTHENING SECURITY OF TRANSACTIONS THROUGH BLOCKCHAIN DB

Blockchain comprises of a record which is disseminated. In blockchain each square has hash key to the past square and rundown of trades. It very well may be a related summary thought which is continually overhauled, imitated and introduced to every one of the hubs. Interlinking between the hubs is finished by hash key, accordingly making a chain of pieces or blockchain [5]. On the off chance that at any second information inside the square changed, hash of the square additionally goes through change. Hash work isn't adequate to neutralize altering. To figure huge number of hashes requires few moments, and this makes block helpless since every one of the hashes of the square can be recalculated. To make blockchain safer Proof of Work (PoW) is utilized. PoW helps in limiting arrangement of new pieces. PoW makes the activity of pieces particularly problematic, since in the event that square is changed, recalculation of PoW is needed for every one of the accompanying squares.

Cloud providers themselves can plot with aggressors for successfully dismissing data respectability. To prevent these plot attacks and to avoid stun trust on the uprightness ensures insisted by cloud suppliers, we propose a creative technique for abusing the square chain development to plan and execute an appropriate, secure database which is blockchain-based for disseminated processing conditions [6]. The main aim is to deploy a cloud-based BaaS which is scalable.

8.3 BIOMETRIC SOFTWARE AS A SERVICE (BAAS)

The main aim of biometric system is to capture the biometric trait, preprocess it and extract the Feature Vector (FV) from the trait. The FV is then stored onto a database and then this data will be used for training and testing of the system.

SaaS is a cloud model which allows user to deploy their services on the cloud infrastructure without a need to worry about the hardware, operating system, security and other concerns [6]. This makes the application development process agile and scalable. The biometric recognition system is designed as a service running on the cloud, and its Application Programming Interface (API) interface will be used by the frontend applications for the consumption of the service.

8.4 BAAS AND CLOUD BIOMETRICS

The recent systems have been widely adapting cloud as it provides a scalable environment for ever-increasing user base which is having laptops, desktops and smartphones with sensors capable of capturing biometric data [7,8]. The BaaS deployed on cloud is referred as biometrics in the cloud [9].

Additionally, Bommagani et al. [10] showed a framework for face biometrics which is cloud based underlining the parallelization of affirmation tasks over various workers. Absolutely these consider pointed streamlining the biometric strategies by scattering the duties. Changed with such cloud-based approaches, the progressing biometric plans have perceived the advantages of biometrics-in-the-cloud plan. At the point when such an arrangement is advanced by a help provider, by then it is insinuated as BaaS.

The later BaaS courses of action as a general rule center around a very much described, however, unassuming degree, for instance giving administrations of a unimodal biometric confirmation structure over cloud. In any case, there's a developing need on using multimodal biometrics [11] and covering too enlarged challenges and use cases with respect to biometrics. A particularly multidimensional technique requires a broader and sweeping comprehension of biometric thoughts, through an ontological view, inside trendy BaaS that can deal with a wide extent of issues.

8.5 EXISTING WORK

In the paper "Online Signature Recognition Using Software as a Service (SaaS)", the authors have discussed a model on Microsoft Azure Cloud [12]. In this model, a Biometric SaaS is running on a public cloud structure like Microsoft Azure; the organization and the data store are on the cloud. The authors in the paper "Biometric Authentication as a Service (BaaS): a NOSQL Database and CUDA Based Implementation" have proposed the scalable and faster biometric system exploiting power of Graphical Processing Unit (GPU). At the backend the FVs and user data are stored on a Column Family Database which brings scalability and the processing is done on a NVIDIA CUDA procedure; this makes the BaaS faster and scalable. In the research paper [12], "Biometric Transcendentalism for Semantic Biometric-as a-Administration. BaaS Applications: Out Skirt Security Use Case", the authors have discussed the solid need on developing a semantic construction that should rely upon a biometric cosmology. The Baas on cloud has five stages of execution, and they are explained as follows.

8.5.1 Biometric Trait Capture and Preprocessing

For online signature recognition system under consideration a digitizing tablet, pen tablet or smart pen is used to get the data for signature. Wacom Intuos 4 Digitizer is used for taking modernized signature; this interfaced through a VBTABLET Digitizer interface through .NET structure. The caught mark is preprocessed to eliminate examining blunders.

8.5.2 Extraction of FVs

The signature-based biometric system uses the dynamic data of handwritten signatures. The data such as X, Y and Z coordinates of the pen tip along with Pressure, Azimuth and Altitude information is captured for each FV. The velocity in X and Y directions, and Acceleration information are extracted.

8.5.3 Matching

The FV extracted is an N-dimensional vector in a Hyper space. These FVs are passed to classifiers for learning purpose and then classifier evaluates the distance between the FVs. This distance is then used for matching the FVs.

8.5.4 Decision

At the point when a similarity score is calculated, the decision recommends the figuring of a decision limit. On the off chance that the similarity is bigger than a limit, the decision is recognized as real; else it is excused as false.

8.5.5 Classification

The signature recognition system is designed for classification of the incoming biometric trait into Accepted and Rejected, the features are extracted and matched and the score is calculated. These FVs are used in such manner that it can coordinate with the classification. The indistinguishable scores are used for user group formation. K Nearest Neighborhood (KNN) classifier is used in this work.

Deployment of the BaaS Design for IoBT

8.6 MODIFICATION OF EXISTING SYSTEM

The current deployment is proposed as a cloud-based version as depicted in Figure 8.1 [13]. In this system, a Biometric SaaS is being deployed on Amazon AWS or Microsoft Azure Cloud. The administration other than information stockpiling is on the cloud. The signatures are captured by Digitizer tablet. Then the captured biometric traits are uploaded on the AWS S3 or Microsoft Azure Datastore. The Event Hub and Apache Storm service manage the incoming captured stream of the biometric traits. Apache HBase and Hadoop provide the distributed feature extraction from the dynamic signature samples (Figure 8.2).

Key events are trait capture, uploading of the trait from the client, feature extraction, storage of the FVs, matching and classification. The above-mentioned system is deployed on Microsoft Azure Cloud.

8.7 BAAS DEPLOYMENT ON AMAZON AWS CLOUD

In this section the AWS-based deployment of the online signature recognition system is discussed. Figure 8.3 shows the architecture. Here the FV is extracted on NVIDIA GPU using a suitable EC2 instance. The EC2 GPU compute service is used for the same. The signature templates are stored on the Simple Storage Service–based buckets. A NOSQL-based DynamoDB is used for the storage of the FVs. NOSQL database are highly scalable and fast for cloud-based deployments. The CUDA GPUs provide parallel execution of the FV extraction mechanism. The code for the same is accordingly modified here for the CUDA Compatibility.

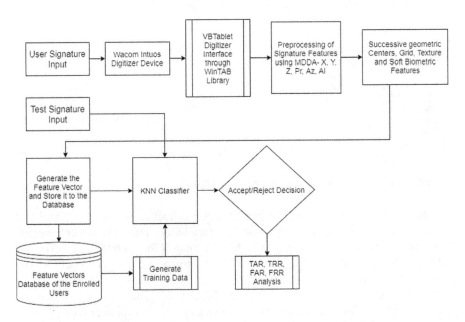

FIGURE 8.1 Design details for the current biometrics system.

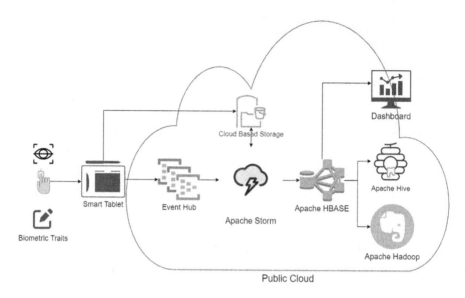

FIGURE 8.2 Engineering of biometric system on cloud network [13].

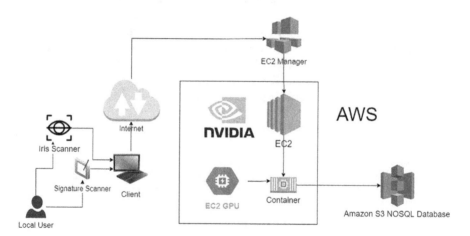

Biometric Authentication as a Service (BaaS)

FIGURE 8.3 BaaS deployment on the Amazon AWS [14].

The NVIDIA CUDA-based component vector extraction is considered here, and Table 8.2 shows the results; the dynamic signature-based systems with soft biometric features have better performance, and NVIDIA Titan X Pascal GPU has overpassed CPU and NVIDIA 1050 GPU. More than 300% timing improvement is achieved by GPU-based systems.

To build the IoBT we need to set up the biometric recognition scripts running on a cloud instance. The following section shows a stepwise description to run a Python script on an AWS EC2 instance.

8.8 IoBT BACKEND

For the cloud-based demo a multimodal system consisting of signature and iris recognition is considered; out of the same iris recognition is tested on both local and cloud-based deployment. This forms the cloud-based backend for the IoBT. The concerned iris recognition was initially implemented in C# and later a Python code was obtained. The Python-based program is deployed on Amazon AWS Cloud, and the deployment steps are as follows. An EC2 instance-based deployment is discussed here. Since you have an AWS account, you need to dispatch an EC2 virtual worker case on which you can run Keras [15]. Dispatching a case is as simple as choosing the picture to stack and starting the virtual worker. The Deep Learning Image will be utilized for the arrangement.

8.8.1 Step 1: Login to Your AWS Console and Create Instance

Launching a new virtual server so click on EC2. Click the Launch Instance button (Figures 8.4 and 8.5).

An AMI is an Amazon Machine Image selected for the code deployment (Figure 8.6).

Select the g2.2×large machine. This incorporates a GPU that can be used to fundamentally increase the execution speed (an indicative screenshot is given below) (Figure 8.7).

Select Review and Launch to finish the setup of your worker occasion, and click the Launch button. Select the key pair. If you have a key pair because of prior utilization of EC2 occasion, select a current key pair from the rundown. On the off chance that you don't have a key pair, select the choice Create another key, combine and enter a key pair name, for example, iris recognition. At that point download key pair.

In the terminal change path to the location of your key pair. If you have not effectively done as such, limit the entrance authorizations on your key pair file.

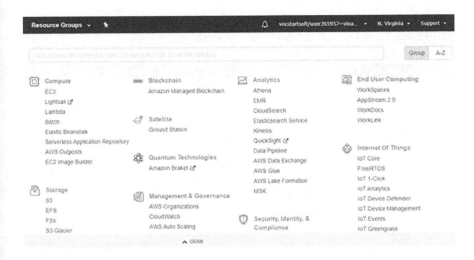

FIGURE 8.4 Amazon AWS console home.

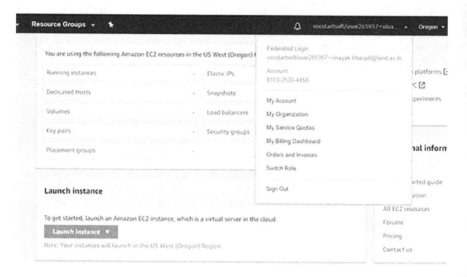

FIGURE 8.5 Launch instance on Amazon AWS console.

FIGURE 8.6 Amazon machine images for deep learning.

FIGURE 8.7 Select required hardware.

Deployment of the BaaS Design for IoBT

This is needed as a feature of the SSH admittance to your worker. For instance, open a terminal on the workstation and type (Figure 8.8):

```
$$ cd Downloads
$$ chmod 600 dynamicRecognition.pem
```

Click Launch Instances. Amazon will approve your solicitation, and this may take as long as 2 hours (for the most part it is only a couple minutes). The worker is presently running and prepared to be signed in as shown below (Figure 8.9).

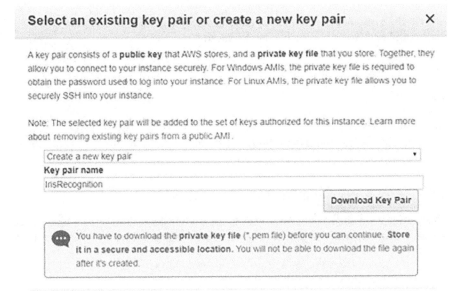

FIGURE 8.8 Select/generate key pair.

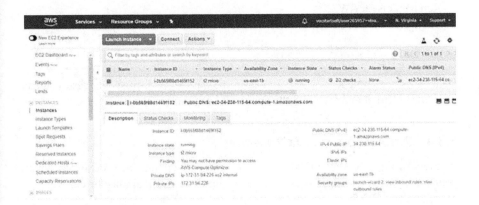

FIGURE 8.9 AMAZON AWS EC2 instance.

8.8.2 Step 2: Login, Configure and Run

Snap View Instances in your Amazon EC2 support on the off chance that you have not done so as of now. Save the Public Internet Protocol address to the machine's clipboard. In this model the IP address is assigned as 34.238.115.64 (please don't use this IP address; it won't fill in as your specialist IP address will be one of a kind). On the terminal window modify the list path to the key pair location.

```
$$ ssh -i "IrisRecognition.pem" ec3-user@ec2-34-238-115-64.
compute-1.amazonaws.com
```

You are currently signed in to your worker.
   This EC2 instance needs small updates and they will be done as follows.
   The updated Keras version can be displayed by

```
$$ python -c "import keras; print(keras.__version__)"
```

You should see:

```
$$ <module 'tensorflow._api.v2.version' from '/usr/local/lib/
python3.6/dist-packages/tensorflow/_api/v2/version/__init__.py'>
2.3.1
```

8.8.3 Step 3: Build and Run Models on AWS

This is initiated by duplicating the records to the AWS instance. The explicit source code registry can be duplicated to your AWS instance utilizing the SC command as given next:

```
$$ scp -iirisRecognition.pem -r srcec2-user@ec2-34-238-115-64.
compute-1.amazonaws.com:~/
```

This will copy the complete src/ directory to the AWS instance's home directory. This example can be modified to get the largest datasets present on the workstation onto the AWS ec2 instance; this activity might be chargeable by AWS.
   Run Models on AWS
   The Python script can be run now on the AWS instance as follows:

```
>> python irisRecognition.py
```

Hence it is recommended to run scripts in the background. In such a case you can close the terminal and the AWS instance will keep the script running.

8.8.4 Step 4: Close Your EC2 Instance

As every instance is billed in Pay as you go mode, at the point when you are done with your work you should close your occurrence. First the terminal must be closed by

```
>>exit
```

Deployment of the BaaS Design for IoBT

```
    _|  _|_  )
   _| (  ~  /   Deep Learning AMI for Amazon Linux
  __|\___|__|

The README file for the AMI  ++++++++++++++++++  /home/ec2-user/src/README.md
Tests for deep learning frameworks  ++++++++++++  /home/ec2-user/src/bin
```

FIGURE 8.10 AMAZON AWS EC2 instance login screen.

Go to EC2 dashboard as shown in Figure 8.10. Select the running instance from the list, select Instance State and choose Terminate instance option. Confirm the same. After some time, the instance state will be indicated as terminated, ensure the same and now you can log out of AWS.

The above-mentioned process will launch an iris recognition Python script which will do the job of iris recognition on an AWS instance, this is an example of BaaS, this forms a cloud-based backend for the iris recognition system implemented as IoBT, and this will accept iris scans from an IoT network and will perform authentication based on the enrolled users data.

8.9 PROPOSED SYSTEM AND INITIAL DEPLOYMENT RESULTS

The proposed cloud-based scalable framework as demonstrated in Figure 8.11 is conveyed on cloud alongside GPU [16–19]. The biometric authentication transactions and the FVs are stored on blockchain-based database for further security and non-repudiation. Key points for the proposed systems are:

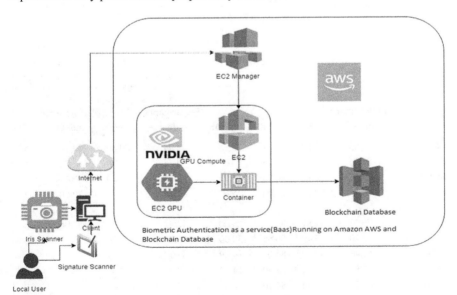

FIGURE 8.11 Deployment of BaaS on AWS and blockchain DB.

1. A multimodal biometric authentication deployed on AWS Cloud as BaaS [4,8,15].
2. Developing faster cloud application for GPU and hybrid hardware [12].
3. Scalable applications with better resource provisioning [20–23].
4. Modification of the design for the incorporation of IoBT. To provide security, accountability for the application access [8–11].
5. Enforcement of provenance property through the use of blockchain database with the proposed frameworks [13–15]. This mechanism is aimed at the intra and inter class division (Table 8.1).

The proposed methods were tested for both online as well as offline biometrics. The geometric center-based FV extraction for offline signature recognition has given best results of 88.3% of PI. Further the work is extended for online signature recognition with its FV extraction on cloud. The GPU- and CPU-based program runtimes are compared and shown in Table 8.2.

For the CUDA-based deployment [14], Table 8.2 shows the outcomes, Dynamic CUDA execution giving better results and that NVIDIA TXP GPU has preferable execution over CPU and NVIDIA1050GPU. The NVIDIA GPU gives over 300% expansion in the program runtime. Table 8.2 shows GPU-based usage subtleties and

TABLE 8.1
Performance Comparison for the Existing Systems [1,13,15]

	Metrics of Performance							
	Exclusive Feature Vector		Unique Feature Vector Joined with Soft Biometrics Features		Improvement in PI & SPI in view of SBF Union (%)		System Performance in Offline Mode	
Feature Vector	PI (Performance Index)	SPI	PI- SBF	SPI-SBF	PI (Performance Index)	SPI	PI (Performance Index)	SPI
GCH	88.33	9.09	91.37	18.18	3.44	100.00	83.14	5.29
GCV	87.14	11.11	91.18	20.00	4.63	80.00	81.22	7.9
GCH+GCV	87.69	10.0	90.96	16.95	3.73	69.50	NA	NA
Grid Features	69.25	32.00	85.03	25.00	22.79	−21.88	82.00	02.10
TXC	75.73	14.81	85.14	25.00	12.43	68.75	81.00	06.10
TFD1	82.70	15.79	85.55	30.00	3.45	90.00	-	-
TFD2	60.43	40.00	86.03	14.29	42.37	−64.29	-	-
TFD3	55.84	14.29	85.40	14.29	52.95	0.00	-	-
TFD4	63.95	5.24	84.70	14.29	32.44	172.73	-	-
TFD5	71.29	14.29	84.44	14.29	18.45	0.0	-	-

TABLE 8.2
Program Runtime (CPU as well as GPU)

Biometrics Specification	CPU (Milliseconds)- Based Execution [7]	NVIDIA GTX GeForce 1050 (Static)	NVIDIA GTX GeForce 1050 (Dynamic)	NVIDIA GTX GeForce Titan X Pascal (static)	NVIDIA GTX GeForce Titan X Pascal (Dynamic)
Signature	17.00	6	5	5	5
Iris right	10	4	4	3	3
Iris left	10	4	4	3	3
Sign + iris L	32	11	11	10	9
Sign + iris R	30	10	10	9	8
Sign + iris L + R	40	14	13	13	12
For 10 sign	200	65	64	62	61
For 10 iris	250	80	78	75	74

FV extraction on NVIDIA GPU reduced the code runtime by 60%. The existing technique was tried on 108 clients, and complete 2.7 K Genuine (Intra Class) and 290 K (Inter Class) Tests were finished. In any case, the above design isn't adaptable. At that point, we move to the scalable Biometric Architecture framework on cloud as discussed in Section 8.8.

8.10 CONCLUSION

In this chapter, Biometric framework, Framework of Biometric System on Cloud and Design of BaaS on Amazon AWS are discussed. The static and dynamic signature recognition systems (online and offline) are discussed. The performance comparison of the proposed systems on CPU as well as GPU is presented here. The final performance in TAR (Performance Index) is achieved up to 91.4%. The soft biometric feature has improved the TAR by 3%–52%.

To make the system scalable and fast, cloud-based systems are explored, and the GPU-based deployments are found to be about 200% faster. To make the transactions accountable and secure use of blockchain DB is also proposed here and an architecture based on Amazon AWS is also discussed.

ACKNOWLEDGMENTS

The GPU-based testing is done on a NVIDA GPU, Titan X Pascal. This GPU was provided by NVIDIA, Inc. under NVIDIA GPU Grants program.

A part of this research work is supported by Amazon AWS Educators Grant. Amazon AWS-based deployment was tested on the AWS Cloud, and access was provided as a part of this grant.

REFERENCES

1. Kekre, H. B.; Bharadi, V. A.; Dynamic signature pre-processing by modified digital difference analyzer algorithm, *Proceedings of Springer International Conference ThinkQuest*, 2010, Springer, New Delhi, Doi: 10.1007/978-81-8489-989-4_12.
2. Jain, A. K.; Flynn, P.; Ross, A.; *Handbook of Biometrics*, Springer, ISBN-13: 978-0-387-71040-2, pp. 1–23, 2007.
3. Jain, A. K.; Ross, A.; Prabhakar, S.; An introduction to biometric recognition, *IEEE Transactions on Circuits and Systems for Video Technology*, 2004, *14*, 1.
4. Kanak, A.; Biometric ontology for semantic biometric-as-a-service (BaaS) applications: a border security use case, *IET Biometrics*, 2018, *7*, 6, 510–518, Doi: 10.1049/iet-bmt.2018.5067.
5. Sharma, S. G.; Ahuja L.; Goyal, D. P.; Building secure infrastructure for cloud computing using blockchain, *2018 Second International Conference on Intelligent Computing and Control Systems (ICICCS)*, Madurai, India, 2018, 1985–1988, Doi: 10.1109/ICCONS.2018.8663145.
6. Armando, A.; Baldoni, R.; Focardi. R.; Blockchain-based database to ensure data integrity in cloud computing environments, *Proceedings of the First Italian Conference on Cybersecurity (ITASEC17)*, Venice, Italy, 2017.
7. Castiglione, A.; Choo, K.; Nappi, M.; Narducci, F.; Biometrics in the cloud: challenges and research opportunities, *in IEEE Cloud Computing*, 2017, *4*, 4, 12–17, Doi: 10.1109/MCC.2017.3791012.
8. Talreja, V.; Ferrett, T.; Valenti, M.C.; Biometrics-as-a-service: a framework to promote innovative biometric recognition in the cloud, *arXiv preprint arXiv:1710.09183*, 2017.
9. Stojmenovic, M.; Mobile cloud computing for biometric applications, *2012 15th International Conference on Network-Based Information Systems*, Melbourne, VIC, 2012, 654–659, Doi: 10.1109/NBiS.2012.147.
10. Bommagani, A; ValentiM.; Ross, A.; A framework for secure cloud-empowered mobile biometrics, *2014 IEEE Military Communications Conference*, Baltimore, MD, 2014, 255–261, Doi: 10.1109/MILCOM.2014.47.
11. Ross, A.; Jain, A.; Multimodal biometrics: an overview, *2004 12th European Signal Processing Conference*, Vienna, 2004, 1221–1224. ISBN: 978-320-0001-65-7.
12. Sahl, R.; Dupont, P.; Messager, C.; Honnorat, M.; La, T.; High-resolution ocean winds: hybrid-cloud infrastructure for satellite imagery processing, *2018 IEEE 11th International Conference on Cloud Computing (CLOUD)*, San Francisco, CA, 2018, 883–886. Doi: 10.1109/CLOUD.2018.00127.
13. Dsilva, G, Bharadi, V.; Biometric authentication using Software as a Service (SaaS) Architecture with real-time insights, *2016 International Conference on Computing Communication Control and automation (ICCUBEA)*, Pune, India, 2016, 1–6. Doi: 10.1109/ICCUBEA.2016.7859980.
14. Bharadi, V.; Mestry, H.; Watve, A.; Biometric Authentication as a Service (BaaS): a NOSQL database and CUDA based implementation, *5th IEEE International Conference ICCUBEA 2019*, Pune, India (Under Publication), ISBN: 978-1-7281-4042-1/19.
15. Bharadi, V.; DSilva, G.; Online signature recognition using Software as a Service (SaaS) model on public cloud, *2015 IEEE International Conference on Computing Communication Control and Automation*, Pune, 2015, 65–72, Doi: 10.1109/ICCUBEA.2015.208.
16. Yadav, A. K.; Bharti, R.; Raw R. S.; SA^2-*MCD*: secured architecture for allocation of virtual machine in multitenant cloud databases, *Big Data Research: An International Journal*, 2021, *24*, Doi: 10.1016/j.bdr.2021.100187.

17. Yadav, A. K.; Bharti, R.; Raw, R. S.; Security solution to prevent data leakage over multitenant cloud infrastructure, *International Journal of Pure and Applied Mathematics*, 2018, *118*, 07, 269–276.
18. Kumar, M.; Yadav, A. K.; Raw, R. S.; Global host allocation policy for virtual machine in cloud computing, *International Journal of Information Technology*, 2018, *10*, 279–287.
19. Ahmed, A.; Abdul Hanan, A.; Omprakash, K.; Lobiyal, D. K.; Raw, R. S. Cloud computing in VANETs: architecture, taxonomy and challenges, *IETE Technical Review*, 2017, *35*, 5, 523–547.
20. Shwe, H. Y.; Chong, P. H.; Scalable distributed cloud data storage service for internet of things, *IEEE International Conferences on Ubiquitous Intelligence & Computing, Advanced and Trusted Computing, Scalable Computing and Communications, Cloud and Big Data Computing, Internet of People, and Smart World Congress*, 2016, Doi: 10.1109/uic-atc-scalcom-cbdcom-iop-smartworld.2016.0137.
21. Srirama, S. N.; Iurii, T.; Viil, J.; Dynamic deployment and auto-scaling enterprise applications on the heterogeneous cloud, *2016 IEEE 9th International Conference on Cloud Computing (CLOUD)*, 2016, Doi: 10.1109/cloud.2016.0138.
22. Kiss, T.; Scalable multi-cloud platform to support industry and scientific applications, *2018 41st International Convention on Information and Communication Technology, Electronics and Microelectronics (MIPRO), Opatija*, 2018, 0150–0154. Doi: 10.23919/MIPRO.2018.8400029.
23. KERAS Deep Learning Framework - https://keras.io/why-use-keras/.

9 A Comprehensive Survey of Geographical Routing in Multi-hop Wireless Networks

Allam Balaram
MLR Institute of Technology

Manda Silparaj
ACE Engineering College

Shaik Abdul Nabi
Sreyas Institute of Engineering and Technology

P. Chandana
Vignan Institute of Technology and Science

CONTENTS

9.1 Introduction: An Overview 142
    9.1.1 Challenges Related to Mobility in Multi-Hop Wireless Networks 144
    9.1.2 Simulator Support for Mobility Models in Multi-hop Wireless Networks 146
9.2 Various Routing Protocols Applied for MWNs, MANETs, VANET, WSN 148
    9.2.1 Geographical Routing Protocols for MWNs 148
        9.2.1.1 Classification of Geographic Routing 149
        9.2.1.2 Greedy-Based Routing 149
        9.2.1.3 Face Routing 150
        9.2.1.4 GFG Routing 150
        9.2.1.5 Opportunistic Routing 151
        9.2.1.6 Void Handling in Geographical Routing 151
    9.2.2 Geographical Routing in MANET 152
        9.2.2.1 Geographical Routing in Aeronautical Ad-hoc Network (AANET) 155
    9.2.3 Geographical Routing in WSN 155
        9.2.3.1 Geographical Routing in Underwater Wireless Sensor Network (UWSN) 155

DOI: 10.1201/9781003206453-9

9.2.3.2 Geographical Routing in VANET 158
 9.2.3.3 Geographical Routing in DTN 161
9.3 Future Work and Research Challenges.. 161
9.4 Conclusion 163
References.. 164

9.1 INTRODUCTION: AN OVERVIEW

Recently, Multi-hop Wireless Networks (MWNs) have become very popular and receive a great deal of attention in research, due to its easy and low-cost deployment. In MWNs, the wireless nodes form a network in a shared wireless medium, with or without the aid of any infrastructure. Multi-hop routing has become an essential component of MWNs due to the short communication range of mobile nodes. Thus, the intermediate nodes act as routers between the source and destination to establish communication. The data packets are transmitted in a hop-by-hop manner with the co-operation of intermediate nodes. Mobile Ad-hoc NETwork (MANET) is the theoretical foundation of MWN that fuels extensive research activity to expand the capabilities of wireless networks and potential applications [1]. This massive expansion significantly contributed to the development of novel solutions in real multi-hop network environments such as wireless mesh network, opportunistic network, delay-tolerant network, flying ad-hoc network, wireless sensor network, underwater sensor network, wireless multimedia sensor network, and vehicular ad-hoc network. This survey mainly concentrates on geographical routing [2] and its mobility issues in prominent MWNs such as MANET [3], Flying Ad-hoc NETworks (FANETs) [4], Wireless Sensor Network (WSN) [5], Vehicular Ad-hoc NETwork (VANET) [6–9], and extension of these networks that evolved significantly due to its emerging applications and recent research attention. The geographical routing has become an essential solution for information delivery to overcome the scalability and mobility issues in MWNs [10–12]. Compared to topology-based routing, geographic routing reduces delay and substantial routing overhead when large networks operate with highly dynamic mobility. Geographical routing eliminates the use of expensive control packets and enables only the next-hop link toward the destination. If the source knows the destination location, a network-wide search for the destination by sending the control packets is eliminated. The geographical routing takes the packet forwarding decision purely based on the physical location information and enables an efficient context-aware routing. To decide the next hop, each node must be aware of the physical location of neighbors, and thus it requires efficient location service and location update scheme. The distance between the mobile nodes is estimated by the signal strength of the received packet or time delays in direct communication. In an alternative way, a node can obtain the location information, by directly communicating with a satellite using a Global Positioning System (GPS), if the mobile node is equipped with a GPS receiver. Recently, the position-based routing approach becomes practical due to the fast development of software and hardware solutions in the determination of relative coordinates. The extensive use of GPS-enabled mobile devices and the scalability of geographic routing increase the usage of location information in MWNs. However, the geographic routing needs to update the location information frequently depends on the node mobility, in turn, high mobility increases the rate of a location update.

Node mobility is an essential characteristic of MWNs that create a significant impact on multi-hop routing protocols [13]. Mobility induces frequent path breakages in multi-hop that lead to increasing the overhead and delay. Thus, multi-hop routing has to adapt to the mobility pattern of nodes to deliver the performance. To enhance performance, understanding the movement of the mobile node is essential. Mobility models are designed with the support of mobility traces collected from humans, wild animals, vehicles, and other real-world objects. As mobility has become an integral part of the wireless network, a wide range of mobility models have been proposed recently to capture the realistic mobility of objects depending on the applications. As everything is moving in real environments, mobility is a crucial concept in the wireless networks that describe the reality of node mobility. Although the geographical protocol need not store the network topology information, the nodes must still be aware of where one-hop neighbor nodes which are physically located in the network. The negative impact of node mobility is significantly high on the maintenance of accurate neighbor lists. The node mobility also improves the coverage of wireless sensor networks [14] and also supports to improve the capacity and security of MANETs [15,16]. However, there is a discrepancy between movement patterns observable in reality and the derived theoretic mobility models for wireless networks. It is essential to prove the efficiency of the dynamic location update scheme for the derived mobility models to attain highly reliable geographic routing protocol. In geographical routing, location information is paramount for routing decisions, and the accuracy of location information has a massive impact on routing performance. Location error or inaccuracy is often combined with the problems caused due to unpredictable node mobility [17,18]. Node mobility is considered as the primary factor in the location update scheme, and it degrades the probability of accuracy of neighbor nodes discover process. The essential factors that impact the geographical routing are the freshness of location information and adaptive location update scheme with the characteristics of node mobility models. The mobility model provides the pattern of mobility in terms of node position, speed, and direction over a specific period [19]. It describes the characteristics of mobile nodes in the dynamic network topology [20]. The inaccurate location information due to the unpredictable node mobility and its pattern decreases the geographic routing performance. It is because the time gap between the location update beaconing is larger than the time of employing the location information for packet routing. It increases the importance of characterization of node mobility and its models with geographic routing. Initially, the proactive beaconing approach is employed for maintaining the neighboring node information. Each node frequently updates the location cache table using beacon packets, but it induces high routing overhead. In case of high interval time in location update, it may lead to outdated position knowledge or location inaccuracy under a highly dynamic network topology. Recently, the design of the location update scheme has considered the mobility parameters such as speed, direction, prediction, and time for observing node location and improves the performance of geographical routing under wireless multi-hop environment.

To identify the destination location and the next-hop node to reach the destination, each node needs a location service and location update scheme, respectively [21]. Each node informs its location to the one-hop neighboring node by sending beacon packets according to the location update scheme. These features pose significant challenges in the research area [22], and it is essential to describe the impact of different mobility models on location services over geographic routing. Designing a new mobility adaptive location service with minimum overhead is essential. It is clear that the performance of the geographical routing purely depends on the accuracy and efficiency of location update that relies on the node mobility and their pattern. The node mobility effect on geographic routing is different when it emulates different mobility model. Recently, the mobility prediction scheme is combined with location services. However, it is not possible for different mobility models that exclude the uniform pause time. With the realization of the importance of node mobility in geographical routing, many works have been proposed to characterize the mobile nodes and routing with respect to the mobility model. It is essential to consider mobility characterization and randomness in the design of location services for enhancing geographical routing.

9.1.1 Challenges Related to Mobility in Multi-Hop Wireless Networks

The performance of MWN is affected in different ways by mobility. The mobility models for diverse wireless networks have some common drawbacks [13,17,18]. These are listed as follows:

1. The rapid advancement of ICT enables mobility and mobile devices as an inevitable part of life. Developing a mobility model that mimics the real movement of mobile objects proved a significant challenge while testing protocols using the simulators. It is also challenging to generalize real-time applications from small-scale simulations. Other factors are an irregular movement of the human that creates a significant challenge in movement prediction, and lack of migration of wireless devices out of a modeled area as the movement of humans cannot be restricted in real time.
2. Only a few mobility models consider the obstacles in the simulation. However, in the real world, there are hills, buildings, trees, and so on. However, modeling obstacles only by restricting the communication range is not enough to reflect reality, since the mobile devices should be restricted to move toward the coordinates of an obstacle. The mobile device coverage is not in regular shape, i.e., circle. However, in simulation, the communication range of a wireless device is taken as circular.
3. In the design of the mobility model, poor location prediction accuracy results in the absence of user behavior, community structure, social relationships, and the correlation between user movement.
4. Validating the mobility models is crucial to ensure the applicability of the model to different types of networks. However, it is difficult due to the impracticability of collecting real mobility data.

Traditional geographic routing protocols have been designed for specific applications, and it is working on the assumption that the performance obtained in the simulation is equal to the real time. However, applying geographic routing in different environments further brings the following challenges, as shown in Table 9.1.

TABLE 9.1

Challenges in Modeling the Geographical Routing in Simulation and Real Time

Network	Moment Pattern	Mobility Aspects Impacting Geographic Routing Efficiency	
		In Simulation	In Real Time
MANET	Random	1. Location inaccuracy due to failure of location prediction during mobility 2. Location prediction failure, when the device follows the completely random movement 3. Locating the mobile destination 4. Local maximum problem	1. The mobile nodes location in real-time MANET environment is predictable and not completely random 2. The continuous movement of devices results in unavailability of the router, i.e., local maximum
	Community	1. Lack of social context and its associated group reforming leads to reduced accuracy in prediction 2. Restricted traveling Distance within a modeled area	1. Unrestricted distance to travel 2. Dynamic formation of community
	Simple Human	1. Lack of using the advantage of spatial regularity in human mobility 2. Lack of modeling obstacles during mobility	1. Some nodes meet on a random basis, but some nodes are more frequently and regularly 2. Periodical reappearance at a preferred location 3. Human movement through obstacles is not possible
VANET	Traffic	1. Link availability based intersection selection in routing is affected by the collision 2. Lack of linking driver decision and movement prediction	1. Modeling obstacles in real time are difficult 2. The dynamic driver decision may not always assure the link availability 3. Lack of Road Side Units (RSUs) in real road scenario
	Behavioral	1. Lack of considering the driver decision and speed acceleration together 2. Vehicles random movement on the road increases the frequent beaconing resulting in a collision	1. Vehicle distribution is not entirely random 2. Lack of RSU availability in real road scenario

(Continued)

TABLE 9.1 (*Continued*)
Challenges in Modeling the Geographical Routing in Simulation and Real Time

Network	Moment Pattern	Mobility Aspects Impacting Geographic Routing Efficiency	
		In Simulation	In Real Time
WSN	Controlled	1. No hot spot issue 2. Energy hole only due to the usage in routing	1. Sensor damage by animals or natural disasters 2. Path changes due to obstacles
	Non-Controllable	1. Frequent location error due to mobility reduces the accuracy 2. Sensor movement in the modeled area 3. A high energy consumption	1. The sensor movement cannot be predefined 2. Node mobility reduces the possibility of local maximum and hot spot issues 3. Sensor damage by animals or natural disasters 4. Path change due to obstacles

Mobility modeling plays a vital role in the simulation-based performance evaluation of geographic routing protocols over MWNs. There are several mobility models in MWNs [19]. However, most of them cannot simulate the realistic motion of wireless devices. The gap between the mobility models in a simulation environment and real-time applications impact the evaluation of geographical routing protocols.

9.1.2 SIMULATOR SUPPORT FOR MOBILITY MODELS IN MULTI-HOP WIRELESS NETWORKS

Several geographical routing protocols have been proposed for various application scenarios. The untested new protocols cannot be installed on a large-scale network due to the uncertainty of its performance. They need to be tested with analytical modeling or simulation tools [23]. Analytical modeling faces significant limitations since the deduced results are not precise in terms of the resources consumed. In contrast, affordable simulation tools provide precise results and accurately evaluate the proposed protocols. Simulation allows the designer to get practical feedback and validate complex protocols' correctness and efficiency at an affordable testing cost within a short span of time. The designer can understand the behavior and the complexity of the protocol by applying a different level of abstraction. Table 9.2 lists some of the network simulators and their mobility models.

Several mobility models are developed for various MWNs [19,24]. The conventional research works build communication networks using both experimental and mathematical models to prove the performance of communication networks. In the past decade, communication networks have become too complex for mathematical analysis.

TABLE 9.2
Network Simulators and Mobility Models

Tool	Type	Mobility Models	Language	Opportunities	Network to Model
Bonn Motion	Mobility Generator	Random Walk, Random Way Point, Gauss-Markov model, Manhattan Grid, and Reference Point Group Mobility	Java	Supporting visualization tool	MANET, FANET, WSN, and VANETs
MobiSim		Random Walk, Random Way Point, Random Direction, Gauss-Markov model, Manhattan Grid, Nomadic Community, and Pursue	Java	Modeling traffic signs and traffic lights	MANET, FANET, WSN, and VANETs
NS2	Network Simulator	Random Walk, Random Way Point, Gauss-Markov model, Manhattan Grid, and Reference Point Group Mobility	C++ and OTCL	Visualization is an available and very popular simulator	MANET, FANET, WSN, and VANETs
NS3		Random Walk, Random Way Point, Random Direction, Gauss-Markov model, Manhattan Grid, and Reference Point Group Mobility	C++ and Optional Python Bindings	It supports TCP, UDP, ICMP, IPv4, multicast routing protocols, and CSMA protocols	MANET, FANET, WSN, and VANETs
Omnet++		Random Way Point	C++	Supports a graphical network editor	MANET, FANET, WSN, and VANETs
OPNET		Random Walk, Random Way Point, Random Direction, and Group Mobility	C and C++	Supports very large-scale multi-hop wireless Networks	MANET, FANET, WSN, and VANETs
NetSim		Random Walk and Random Way Point	C and Java	It provides a GUI with the features of drag and drops functionality for devices	MANET, WSN, and VANETs

(Continued)

TABLE 9.2 (*Continued*)
Network Simulators and Mobility Models

Tool	Type	Mobility Models	Language	Opportunities	Network to Model
GloMoSim/ QualNet		Random Walk, Random WayPoint, Gauss-Markov model, Manhattan Grid, Reference Point Group Mobility, and Group Mobility	C	It can scale up to networks with thousands of heterogeneous nodes	MANET, WSN, and VANETs
SSFNet		Random Walk, Random Way Point, Gauss-Markov model, Manhattan Grid, and Reference Point Group Mobility	Java and C++	Scalable, high-performance network modeling	MANET, WSN, and VANETs
J-Sim		Random Way Point	Java	Supports energy modeling, component-based architecture	WSN
GTNets		Random WayPoint	C++		MANET, WSN, and VANETs

Thus, network designers prefer simulation tools for analyzing the behavior and performance of the networks and their protocols. The network simulation tools are often used in testing the capacity of networks to meet the quality of service. Also, the simulation tools can explore a wide range of potential protocols for evaluating wireless network performance.

9.2 VARIOUS ROUTING PROTOCOLS APPLIED FOR MWNs, MANETs, VANET, WSN

9.2.1 GEOGRAPHICAL ROUTING PROTOCOLS FOR MWNs

Recently, MWNs have embraced geographical routing protocols for many potential applications due to the stateless property, scalability, and efficiency. It widely supports energy efficiency, Quality of Service (QoS), node mobility, anonymity and privacy, and context-awareness. Geographic routing algorithms employ position information such as its position, position of neighbors, and destination to make packet forwarding decisions. It delivers data packets in a network over multiple hops utilizing location information. This process makes geographic routing more attractive for dynamic wireless network scenarios. Geographic routing also supports context-aware routing decisions based on location information that leads to the establishment of ubiquitous computing. The geographic routing protocols employ location servers for the sources to obtain the position of the destination. Each node updates its location information to the location servers using a handful of messages. The significant advantage of this kind of routing is that it works nearly stateless and establishes high delivery rates under mobility. The use of geographic routing in wireless networks has tremendously helped in many applications such as rescue operations, military networks, forest fire monitoring, vehicular network monitoring, underwater networks, and flying objects and Unmanned Aerial Vehicle (UAV) networks. This section is dedicated to geographical routing protocols applicable to various MWNs such as MANET, FANET, WSN, Underwater Wireless Sensor Network (UWSN), VANET, and Delay-tolerant networking (DTN). Initially, the section starts with the foundation of geographical routing protocols [25,26].

9.2.1.1 Classification of Geographic Routing

The geographic routing protocol classifies into four significant methods such as greedy-based routing, face routing, greedy-face-greedy (GFG) routing [27–34], and opportunistic routing [35,36], as shown in Figure 9.1. Greedy forwarding is the earliest routing approach in which the data packets are routed to the neighboring node that is closer to the destination. However, greedy forwarding does have one significant problem that is void handling [37]. Compass routing is a basic model for face routing, where faces on a planar graph are traversed using the right-hand rule, kept track of all the time. It successfully detects when a node crosses a line connecting the source and destination [26]. The first face routing algorithm is executed by utilizing the application of the compass algorithm to Unit Disk Graphs (UDG) and for planarizing UDGs [38]. Generally, the UDG is a base model for a planarization algorithm, which is typically based on the Gabriel graph [39].

9.2.1.2 Greedy-Based Routing

The greedy-based routing is a simple, easy to implement, and understandable geographic routing. In greedy routing, when a source wishes to forward the data packets to a destination, it consults the neighbor table to determine a next-hop node that is closer to the destination. Upon receiving the data packet, the next-hop node executes the same procedure to forward the data packets. However, greedy routing does have a significant problem that is when a node has an empty neighbor list to reach the destination; it drops the data packets, referred as local maximum or void problem [40–42]. The local maximum problem provides a severe issue for the performance of greedy forwarding.

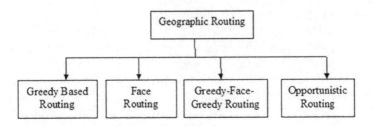

FIGURE 9.1 Classification of geographic routing protocol.

An augmentation of basic greedy routing employs the potential field concept, which is commonly used in robot navigation [43]. The possible fields are used in the movement of robots using the virtual repulsive force. Therefore, the target exerts a positive virtual force, whereas obstacles exert a negative virtual force [44]. The sum of both the positive and negative virtual effects represents the moving direction of the robot. It simulates and demonstrates the protocol performance, which is similar to the Greedy Perimeter Stateless Routing (GPSR) [40]. The greedy forwarding technique is also utilized in wireless 3D networks such as underwater sensor networks and aerial vehicle network with an efficient end-to-end packet delivery ratio.

9.2.1.3 Face Routing

To overcome the problems faced in greedy forwarding, an alternative method called as face routing is applied, which builds planar subgraphs using a polarization mechanism [45]. In the planarization process, the intersecting edges are removed in a network graph for constructing the planar graph and thus avoids segmentation in the network. A UDG consists of a series of circles and a line where two circles are overlapped, and they indicate communication links [38,44]. A UDG can model a mobile ad-hoc network in which the vertices represent the network nodes located on the Euclidean plane. When the transmission range of two nodes is overlapping, they can communicate with each other. An Adaptive Face Routing (AFR) ties the destination determination cost into the routing function. However, due to the limitation of its searching area, the Bounded Face Routing (BFR) with the elliptical shape increases data delay and overhead on routing due to the increased number of hops [46]. The greedy and perimeter routing are combined to enhance geographic routing performance. The UDG model takes into account the collision and noise using Signal Interference plus Noise Ration (SINR) model to modify the UDG in [47]. A stated problem with the SINR model is that it does not provide a decentralized algorithm. Although there are similarities between the combinatorial model of the UDG-SINR model and the Quasi Unit Disk Graph model in [48], the Quasi UDG model is relatively simple, whereas scheduling with the SINR model leads to an NP-complete problem. A stated drawback of the right-hand rule method used in UDG is blind packet traversal, and it leads to a long path and substantial data delay [42]. Due to the claim of incorrect faces in the face routing protocols leads to inefficient routing over highly dynamic environments [45].

9.2.1.4 GFG Routing

The GFG routing protocol or hybrid routing combines both greedy and face routing, where it starts as the greedy routing protocol, and it switches to face routing in the absence of neighbors. Greedy Other Adaptive Face Routing (GOAFR) [49] builds hybrid greedy and face routing. It uses the greedy mode until the local maximum problem is encountered, and OAFR is employed as a recovery scheme for the void problem. Once the face routing, OAFR is initialized and continued until it reaches the destination or the greedy mode is possible to reach the destination. GOAFR (GOAFR+) extension is a hybrid greedy-face routing, and it is asymptotically optimal routing [50]. The main difference between the GOAFR and GOAFR+ is that it does not take the full boundary of a face, instead of using two counters to keep the

A Comprehensive Survey of Geographical Routing

greedy neighbors closer to the destination. For general UDG, a clustered backbone graph is generated using a dominator set, which includes the nodes associated with the backbone in the graph, and the packets are routed to the destination through the ordinary nodes. A Greedy Path Vector Face Routing (GPVFR) is a non-oblivious algorithm that guarantees efficient packet delivery even without complete face information and improves routing performance in terms of both hop stretch and path stretch [51]. GPSR is one of the most widely used geographic routing protocols. The main differences between GPSR and GFG areas follows: the relative neighborhood graph in GPSR changes the face before crossing the range. Another difference is that the time of terminating the face routing and switch back to the greedy mode. In GFG, the face routing is terminated only when the packet arrives at the node, which is closer to the destination. However, in GPSR, the perimeter mode is terminated only when the distance to reach the destination is lower than that of the node where the packet entered into perimeter mode. The work [52] provides a guaranteed delivery and formal proof of many proposed face routing, and combined greedy-face routing schemes.

9.2.1.5 Opportunistic Routing

The nature of broadcasting is the main characteristic of the wireless medium. Multiple nodes can overhear the same packet, even when a node forwards a packet to a specific node [53–57]. Earlier routing protocols treat the broadcast nature of the wireless medium as an inconvenience because it encourages packet collision. A new routing paradigm, named as Opportunistic Routing (OR), exploits the benefits of the broadcasting nature of the wireless medium. Unlike conventional routing protocols, OR selects forwarding candidates instead of choosing a specific packet forwarder. It allows further candidates to store and forward the overheard data packet. The forwarding candidate list is similar to routing table maintenance in conventional routing, but it can be either global or local, depending on the routing strategy. Typically, a source node maintains the global forwarding set, while the local forwarding set is distributed among candidates. All the forwarding candidates need to be coordinated according to some criteria for opportunistically routing the data packets without duplication. The OR protocol transmits a packet through any available links rather than a selected single link, and in other words, additional backup the availability of nodes in OR reduces the node mobility impact on packet delivery dramatically. Therefore, the OR increases the robustness of multi-hop wireless communication.

9.2.1.6 Void Handling in Geographical Routing

Generally, a lot of routing paths are available among a source-destination pair in dense wireless networks. On the contrary, the number of intermediate routers is decreased due to the presence of void areas in sparse networks. Due to the lack of expensive routing path discovery and maintenance, geographical routing is an effective solution in MWNs, although the presence of void areas in the various network areas is a severe limitation in geographic data forwarding. In the presence of a void area, the geographic routing protocols often fail to deliver the data packets at the destination successfully. In such a case, the data packets are dropped in the void area, and further, they are not transmitted due to neighbor unavailability. Thus, the void issue in geographic routing diminishes the performance of MWN and increases unnecessary energy depletion at nodes. A fair routing protocol should attain equal performance in both sparse and dense networks. The existence of the void area is a critical issue, and it is crucial to handle the void areas for successful data delivery. In geographic routing, it is a technical challenge to select the relay nodes in the presence of void areas. Due to the random node topology and dynamic node mobility, detecting the void area is not easy in wireless networks. Without providing appropriate solutions for void handling, the network performance may severely degrade in terms of packet loss and resource consumption. Notably, such behavior is mainly undesirable in applications related to detecting critical events such as fire detection of WSN and collision warning in VANETs. Flooding is the simplest void avoidance method that provides the least routing path for successful data transmission in wireless networks [58,59]. The network nodes receive multiple unnecessary copies for single data transmission. Thus the flooding method is inefficient, as it consumes high energy and bandwidth in resource-constrained networks.

Figure 9.2 depicts the classification of geographic void handling mechanisms. The existing works comprehensively survey the geographic routing void handling techniques by categorizing planar graph-based, geometric, flooding-based, cost-based, heuristic, and hybrid [60]. The planar graph void handling technique utilizes a graph that comprises the not intersect the edges of MWN. In planner graph-based void handling, a right-hand rule-based traversal mechanism is employed for data transmission. The geometric void handling mechanism utilizes the geometric properties of network topology for void handling. The flooding-based techniques employ a simple flooding method to manage void handling issues in data forwarding. The cost-based mechanism uses a shadow and cost spread phases with location information for void handling. The heuristic-based void handling methods utilize some heuristics to manage the void problem. In a hybrid model, two void handling techniques are combined to efficiently forward the data packets.

9.2.2 Geographical Routing in MANET

Geographical routing protocols have significant advantages over topology-based routing in many practical network scenarios in MANETs. The design of geographical routing supports scalability, mobility, and efficiency to manage a wide range

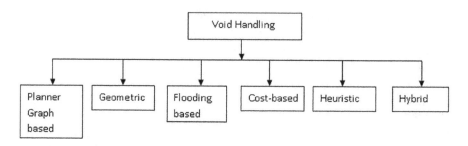

FIGURE 9.2 Classification of geographic void handling techniques.

A Comprehensive Survey of Geographical Routing

of applications in MANETs. This section discusses prominent geographic routing protocols employed in MANETs. The greedy distributed spanning tree routing (GDSTR) [61] utilizes a spanning tree with the planarization algorithms. The spanning tree is also named as a hall tree in which the top or parent nodes maintain the location information about all of their child nodes. As with the backup scheme, the greedy distributed spanning tree routing switches the greedy mode to the spanning tree algorithm, and it assists in handling the local maximum problem. The work in [62] consists of a graph planarization-based backup scheme which uses Rotational Sweep (RS) algorithm, and it is executed when it reaches the local maximum problem. The combination of greedy [63] and contention RS algorithm attains guaranteed packet delivery ratio over connected UDG networks. The RS algorithm uses several hops for data delivery, which is equal to the Gabriel graph planarization algorithms. The counterclockwise sweep based on the work of [64–66] was introduced to find a boundary traversal path [67–73] to escape from the void problem. The main problem associated with the existing backup schemes for escaping from the local maximum problem is that they are initiated only when it reaches the dead end. It makes that the packets need to travel for a long path to reach the normal or greedy mode. Several graphic routing protocols have been proposed for MANETs, normal or greedy mode. Several geographic routing protocols [74–76] have been proposed for MANETs, as illustrated in Table 9.3.

TABLE 9.3
Geographical Routing Protocols in MANET

Routing Protocol	Category	Optimization Objective	Advantages	Limitations
GPSR [40]	Greedy Routing	Maximum progress	Loop free Routing	High Packet Overhead and less scalability
Most Forward within Radius (MFR) [66]		Minimizing the number of Hops	Less Routing overhead	Routing loop may occur
Compass Routing [74]		Positive Progress	Fewer collisions	Delay due to backward communication
Power-Aware [75]		To minimize energy consumption	The better trade-off between delay and energy consumption	Lack of recovery method
GLR [42]		Avoiding unnecessary longer traversal	Avoiding triangular routing reduces the delay	High processing overhead
Normalized ADVance (NADV)-based routing [76]		The optimal tradeoff between proximity and link cost	Efficient and mobility adaptive cost-aware routing strategy	It is inefficient for sparse scenarios.

(Continued)

TABLE 9.3 (*Continued*)
Geographical Routing Protocols in MANET

Routing Protocol	Category	Optimization Objective	Advantages	Limitations
Greedy Routing Protocol with Backtracking (GRB) [63]		To deal with voids and improve routing efficiency	Lower end-to-end delay, overhead, and minimum hop count	Comparable PDR
Energy and Mobility Greedy Perimeter Stateless Routing (EM-GPSR) [67]		Improves a method to deal with routing voids	Less computational complexity	High overhead
UDG [38]	Face Routing	Detour	Generating a planer subgraph for void handling	High delay
GG [39]		Sparse	Closed Polygon boundary ensures data delivery	Scalability issues
Convex [68]		Adaptive face Routing	Scalability	Longer traversal around the planer graph
Original [74]		Reducing the delay of GOAFR	Restricted boundary traversal	High computational complexity
GOAFR [49]	Greedy-Face-Greedy Routing	Adaptive face Routing	Scalability	Full boundary traversal increases the delay
GOAFR+ [69]		Reducing the delay of GOAFR	Restricted boundary traversal	Face routing over non-sparse area increases delay
GFG [51]		Resolving the local maximum issue	Distance-based face termination	Incorrect face changes
Topology Aware Geographic Routing (TAG) [70]		Making better local forwarding decisions and reducing the number of hops	Supporting for node mobility and void handling	High overhead
POR [71]	Opportunistic Routing	Reliable communication	Fault tolerance	Overhearing failure increases Duplication
A theoretical model for OR [72]		Maximum progress	Medium usage	High retransmissions in a high-traffic scenario
geographic routing scheme [73]		Dealing with critical communication voids	High PDR under sparse MANETs	

9.2.2.1 Geographical Routing in Aeronautical Ad-hoc Network (AANET)

AANET is a subset of MANETs [77–83] that comprise a number of aircraft that establish air-to-air communication among them. The AANET includes the ground and satellite networks for establishing multi-hop communication. Due to the smart technological developments, the breed of AANETs receives excellent attention in establishing single or multi-hop communication among aircraft and manages the data flow produced by aircraft for worldwide coverage [84]. The AANETs have a high deviation in velocities, avionics, very dynamic network topology, and constrained resources. Such characteristics make AANET communication is challenging and notably bottlenecks the network performance in low-attitude areas. Due to high aircraft density and very dynamic network topology, the geographic routing protocols are more suitable for AANETs, when compared to topology-based routing protocols. Most of the existing MWN geographic routing protocols exploit the location information that is updated from GPS, and they attain excellent performance in dynamic, varying network topologies. Compared to MANET, FANET, and VANET, the aircraft in an AANET move with very high speed, and it poses unique challenges in routing data packets in AANETs. Therefore, geographic routing protocols are not applied directly to AANET routing. Hence, it is crucial to append the novel aircraft properties with WMN geographic routing techniques for attaining reliable performance. A novel Automatic Dependent Surveillance-Broadcast (ADS-B) offers high rich and precise flight information by exploiting airborne radar. An ADS-B system-aided geographic routing (A-GR) sets up a neighbor table based on location and mobility information obtained using the ADS-B mechanism [85]. The A-GR selects a next hop based on the neighbor table and optimizes the network performance with low overhead.

9.2.3 GEOGRAPHICAL ROUTING IN WSN

Relaying many individual packets that carry similar data to a sink for processing is significantly impacting the scale of the network in terms of energy consumption and latency costs [86–88]. Energy saving is one of the essential features for energy-constrained sensors in WSN. Clustering is essential and plays a vital role in energy saving for WSN [89,90]. Several works design the energy-efficient clustering structure using a mobile sink [91,92]. Organizing the WSN into clusters enables the cluster head to serve as fusion points for aggregating the data to reduce the quantum of data transmitted to the sink in multiple hops. In contrast to the cluster-based routing protocols, the energy-balanced routing algorithm based on geographical information improves reliability and reduces complexity [93]. Diverse geographic routing techniques [94–96] have been proposed and are currently being investigated to make the data collection as efficient as possible. Some of the routing protocols are listed with their advantages and limitations in Table 9.4.

9.2.3.1 Geographical Routing in Underwater Wireless Sensor Network (UWSN)

UWSN is a form of wireless sensor networks, which is a promising method for monitoring oceanic environments [36]. Similar to ground WSN, the UWSN comprises different sensor nodes that collaboratively monitor the specific oceanic environment.

TABLE 9.4
Geographical Routing Protocols over WSN

Routing Protocol	Category	Optimization Objective	Advantages	Limitations
EDCP [94]	Greedy Routing	Energy-Efficient Event Detection	Cluster and greedy model reduce the delay	Poor performance with multiple events
Weighted Bi-partite Matching Protocol (WBMP) [95]		Improving network lifetime	High scalability	Delay is not addressed
Hierarchical Geographic Clustering Protocol (HGCP) [96]		Monitoring environment	Three levels of coordination, improve routing	A static environment
Conical Forwarding Region-based Geographical (CFRG) [97]		Aims to reduce the number of redundant data transmission	Minimizing collision probability in 3D routing	Does not consider overhead
GRACO [27]		Reliable data delivery by adjusting the blocking situation	Avoid void holes and quick data delivery	High data delivery cost
VAA [28]	Face Routing	Recovery algorithm	Distributed Mechanism	High energy consumption
Novel face routing method [29]		Aims to enhance the routing efficiency	Reduce the transmission amount of location information	High computational complexity
SPEED-1	Greedy-Face-Greedy Routing	Void recovery	Planer graph ensures packet delivery	Long delay
SPEED [30]		Efficient Monitoring	2-hop ensures recovery success	High computational complexity
GeoK [31]		Duplication avoidance and void Handling	Reduced delay and energy consumption	High complexity
EXoR [32]	Opportunistic Routing	Maximum progress	Reduced packet duplication	Packet Batch maintenance increases overhead
EEOR [33]		Reducing delay and fast detection	Reduced cost	Scalability issues
MTS [34]		Fast event Observation	Reducing the number of transmissions	Assuming no Duplicate Packet transmission
Congestion-aware opportunistic routing [35]		Congestion avoidance using a sleep scheduling technique	Energy-efficient routing	Medium scalability

A Comprehensive Survey of Geographical Routing

The information gathered from UWSN nodes is forwarded to a static or mobile surface station. Unlike WSN, the nodes in UWSN are mobile, and they approximately move with 2–3 m/sec speed to monitor the various activities of underwater circumstances. Energy limitation is another critical issue in UWSN. Most of the ground WSN routing protocols are not suitable for the USN environment due to high node mobility and energy constraints. Clustering-based geographic routing is an optimal solution for reliable and energy-efficient data delivery in UWSN, especially in a hostile environment [98]. The reason is that the geographic routing protocol does not add any burden to UWSN design. In geographic routing, each sensor node exploits the geographical position information of nodes and selects a next hop closest to destination for data forwarding. However, the dynamic characteristics and three-dimensional deployment of UWSN limit the geographic routing performance. Firstly, the GPS appended with sensors may not work in a marine environment. Hence, it is crucial to utilize sophisticated localization methods to determine the positions of moving sensor nodes. Secondly, the number of hops increases when the sensors are in the deep marine environment. Thirdly, each mobile sensor depletes high energy due to mobility. Finally, the geographic routing mechanism may fail in the presence of a void area [99–101]. Moreover, the ground WSN geographic routing protocols are not directly utilized for UWSN. A lot of new geographic routing mechanisms are introduced for UWSN to overcome such issues. Some of the UWSN geographic routing protocols are evaluated in Table 9.5.

TABLE 9.5
Geographical Routing Protocols in UWSN

Routing Protocol	Category	Optimization Objective	Advantages	Limitations
Vector-based forwarding (VBF) [101]	Greedy Routing	Selecting eligible node with Cartesian routing for packet forwarding	Less energy consumption and reliable router selection	Low packet delivery ratio and sensitivity to routing pipes
Hop-by-Hop VBF (HH-VBF) [102]		Designing hop-by-hop unique virtual pipe routing	Duplicate packet reduction and less energy Consumption	Packet loss due to inefficient desirable factor
Reliable and energy-balanced algorithm routing (REBAR) [103]		To optimize energy consumption, improve packet delivery ratio and handling void area	High scalability and void handling	Lack to handle the void problem
Focused beam routing (FBR) [104]		To attain energy-efficient routing	Suitable for both mobile and static UWSNs	Minimum packet delivery over sparse UWSN due to void area

(Continued)

TABLE 9.5 (Continued)
Geographical Routing Protocols in UWSN

Routing Protocol	Category	Optimization Objective	Advantages	Limitations
Directional flooding based routing (DFR) [105]		Optimized data forwarding strategy by selecting high-quality links	Void handling	Low packet delivery ratio, when no nodes locate closest to sink
Sector-based routing with destination location prediction (SBR-DLP) [106]		Predicts the location of destination for efficient data transmission	Pre-planned movements-based destination location prediction	Medium robustness and does not handle voids
LCAD [107]	Greedy-Face-Greedy Routing	Cluster-based two-level communication	Medium scalability	Medium robustness and no void handling
Stateless opportunistic routing protocol (SORP) [108]	Opportunistic Routing	Aims to detect void area and trapped nodes	Minimum packet loss, energy consumption, and end-to-end delay	Medium scalability
Totally opportunistic routing algorithm (TORA) [109]		Aims to a void horizontal transmission and prolong network lifetime	Void handling and Optimized energy consumption	High overhead

9.2.3.2 Geographical Routing in VANET

The dynamic nature of VANET induces high speed of vehicles resulting in poor performance of traditional topology-based routing protocols of MANETs. They suffer from node mobility since the topology-based routing protocols require to establish the routes in advance and maintain end-to-end communication routes in a routing table. The geographical routing protocols are more suited for highly dynamic network topology, due to the availability of digital maps and GPS receivers in modern vehicles that inspire the use of geographic-based mobility prediction and routing for VANET [21,110,111]. The fundamental GPSR and greedy perimeter coordinator routing techniques transmit data packets by choosing relay nodes toward the destination and employs a store and carry forward mechanism in case of data forwarding fails [112,113]. The fundamental protocols require the network as fully connected for efficient data delivery. Such routing protocols lack to attain high performance under VANET, as the VANET has dynamic network characteristics such as high-speed vehicles, complex road structures, traffic lights, obstacles, and unconnected partitions of network nodes. Some of the conventional geographic routing protocols in VANET are listed in Table 9.6. VANETs employ clustering for improving network

TABLE 9.6
Geographical Routing Protocols over VANETs

Routing Protocol	Category	Traffic Awareness	Advantages	Limitations
GPSR + Advanced Greedy Forwarding (AGF) [112]	Greedy	No	Advanced greedy routing	Local maximum issue
Contention Based Forwarding (CBF) [116]			Beacon-less routing	Medium access delay
Greedy Perimeter Coordinate Routing (GPCR) [113]		Yes	Restricted forwarding reduces the delay	Perform not well in a city environment
Infrastructure Assisted Geo Routing [117]			Less communication delay	Higher cost due to moving RSUs
Greedy Perimeter Coordinator Routing (GPCR) [118]		No	Junction aware router selection reduces the delay	Scalability is not evaluated
Predictive Geographic Routing Protocol (PGRP) [119]		No	Location and angle based router selection	High computational complexity
K-LDTG [120]	Greedy-face-Greedy Routing	No	Restricted boundary traversal reduces the delay	Its failures, when extending it to delay-tolerant applications
The enhanced geographic routing protocol (PGR) [121]		No	High packet delivery ratio with a minimum hop count	Intersection based routing may lead to inefficient path selection
Motion Vector scheme (MoVe) [122]	Opportunistic Routing	No	Suitable for a city environment	Local minimum issue
GeOpps [123]			Velocity vector usage in routing enables fast communication	Lack of coordination among vehicles increases duplication
Static-Node Assisted Adaptive Routing Protocol (SADV) [124]			Reducing the data delivery delay	Not better performance under the high dense scenario

(Continued)

TABLE 9.6 (Continued)
Geographical Routing Protocols over VANETs

Routing Protocol	Category	Traffic Awareness	Advantages	Limitations
probability prediction based reliable (PRO) [125]		Yes	Optimal routing performance due to effective candidate set Selection	No void handling
New routing scheme		Yes	Adaptive for dynamic VANET scenario	High overhead and local maximum problem
Novel opportunistic routing protocol [126]		No	Maximizing the access point connectivity through opportunistic routing	High routing overhead
Time bargaining game based opportunistic routing [127]		Adaptive routing to dynamic VANET Environment	Real-time optimized routing performance	Local maximum issue
3MRP+ [124]	Location prediction	No	Accurate location prediction and best forwarder selection	High computational complexity
Artificial spider-web-based geographic routing (ASGR) [128]		Yes	Minimizing overhead and improving routing efficiency using virtual spiders	No void handling and does not suitable for sparse VANET
Multi-metric geographic routing (M-GEDIR) [129]			Best next hop selection using multi-metrics	Medium robustness and scalability
Connectivity-aware minimum-delay geographic routing (CMGR) [130]			Suitable for both sparse and dense networks	Do not consider overhead
Destination-aware context-based routing protocol [131]		Yes	Enhances VANET performance and reduces delay	Not taking into account the node movements

scalability [114] and group a set of nodes with similar characteristics. Based on clustering rules, the vehicles are divided into different groups [115]. Each cluster includes one leader, named as cluster head, which is responsible for local central management entity and intra-cluster communication. A cluster head is followed by one or more than one cluster member.

9.2.3.3 Geographical Routing in DTN

DTNs [132] are applied to diverse network scenarios such as WSN, UWSN, and sparse VANETs for end-to-end data delivery. The DTNs are mostly exploited for sparse networks. The DTNs suffer from highly varying network topology, and geographic routing is a promising routing that employs real-time position information of nodes instead of topological information. However, the location information of the destination is likely unavailable over sparse WMN in a real-time environment. In such a case, the DTN geographic routing outperforms the topological routing protocols by carrying the history of the location of nodes for effective data delivery. The DTNs maintain fair connectivity between two communicating nodes located within the communication range of each other. Each node in DTN periodically receives a hello message to know its current neighbors and stores the received information for making routing decisions. Each node waits and forwards the messages based on the prediction of future encounter opportunities, which is known as the store and carry forward method. Thus, the DTN nodes handle high mobility and sparse network conditions by asymmetrically forward the messages. The existing geographic routing protocols proposed for WMNs are not feasible for DTNs. The reason is that most of the MWN geographic routing protocols assume the geographic location of the destination, and is always available in the network. Each node makes routing decisions based on the assumed location information, and it may be wrong in the presence of mobile destinations in DTNs. In addition to that, it is challenging to apply geographic routing in DTNs due to the reliability of the information, destination mobility, and void problem. The existing work in [133] comprehensively surveys the DTN routing protocols, and it provides a classification of DTN geographic routing protocols. Some of the convention DTN geographic routing protocols are surveyed in Table 9.7.

9.3 FUTURE WORK AND RESEARCH CHALLENGES

Realistic mobility models are essential to improve the efficiency of geographical routing. The commonly used alternative method is to simulate the geographical routing over realistic movement patterns. Hence, the accuracy of performance evaluation of geographical routing mostly depends on the mobility patterns and its reflection on reality as possible. Human mobility modeling and prediction can significantly assist the geographical routing in emerging MWN applications.

Recently, wireless network applications have emerged in several areas ranging from safety to comfort applications. Thus, it increases the exploration and research on wireless communication in two aspects, including routing and mobility. Node movement plays a vital role in the wireless communication research field with the complexities of scenario characteristics. The future research in this scenario needs to realize that the critical aspect to be considered to attain the network-specific goals in mobility model features. Integrating the realistic mobility model characteristics in geographical routing decisions improves efficiency.

MANET: Mobility patterns have a significant impact on the performance of a routing protocol. The previous research validates that the mobility patterns in a MANET environment are not entirely random. Instead, the random mobility, the

TABLE 9.7
Geographical routing for DTN

Routing Protocol	Network Type	Main Objective	Advantages	Limitations
Vector routing [134]	VANETs	Utilizing the vector of node movements	Improve routing efficiency	Huge overhead
RoRo-LT [135]	Pocket Switched Networks	Spatiotemporal history-based location observation	Replication avoidance	Destination location unawareness decreases the routing efficiency
Motion vector (MOVE) [136]	VANETs	Node moving direction-based routing	Enhanced routing performance	Fail to address the local maximum issue
GeOpps [137]	VANETs	Optimal router selection based on METD	Low delay in data delivery	Scalability issue and no void handling
GeoSpray [138]	VANETs	High routing efficiency and effective resource utilization	Duplication avoidance in packet delivery	High computational complexity and no void handling
AeroRP [139]	Airborne Networks	Distance and moving direction-based router selection	Optimal performance under high dense networks	Not suitable for sparse network and not considering the local maximum Issue
Delegation geographic routing (DGR) [140]	VANETs	Reduce false routing decisions due to inefficient moving direction	Solves the local maximum problem	Lack to avoid message duplication

devices that are carried by humans are predictable. In conventional, several works attempt to model human mobility. Even though the solutions are developed for understanding the movement patterns, it is essential to change the trends in the human mobility model by considering the obstacles, especially in geographical constraints, while crossing the obstacles. It also leads the routing protocol to several changes in MANET.

FANET: Designing good mobility models by integrating smooth characteristics of FANETs is fundamental ad-hoc mobility models in order to reflect the realistic flight behavior [141,142]. Also, it is necessary to provide an in-depth analysis of UAV movement patterns in a real-time environment. Another future work is to adaptive localization mechanisms for improving geographic routing efficiency.

WSN: Modeling macro mobility in WSN is a significant challenge to cope with the constraints in the sensor network, such as low battery power capabilities. Energy-efficient data collection and mobility management in WSN demand great efforts in the analysis of mobile sinks. The mobility model has to reflect various scenarios depending on the applications such as hole due to dead nodes, and damage of sensors. The mobility model for sensor networks deployed on animals that move very slowly and sometimes stay for a long time at the same location; the speed of the motion element must accommodate the differences. However, in running cases, speed will be increased.

VANET: The importance of vehicle movement in reflecting and inducing a realistic road environment in simulation necessitates the design and application of realistic mobility models in the core of geographical routing [143–145]. Some of the conventional mobility models rely on micro-mobility aspects such as vehicle speed, direction, and geographical constraint on a road, and other models focus on the sociological aspects. In order to support a more realistic mobility model and adapt the simulation environment to specific environments, it is imperative to focus the concepts of micro and macro mobility factors jointly. Integrating Internet of Things (IoT) with VANET creates flexible future enhancement in Intelligent Transportation System (ITS) and also offers real-time responses to the drivers.

Machine Learning and Artificial Intelligence: Location and mobility prediction by applying Machine learning (ML) and Artificial Intelligence (AI) can significantly improve the performance of routing in a highly dynamic pervasive MWN environment.

IoT: In recent years, IoT plays a vital role in several MWN applications such as healthcare automation, industrial monitoring, smart city, smart transportation, and agriculture. Generally, the IoT sensor nodes collect the data about the monitoring environment and forward the data with the help of external internet services. Thus, the IoT enhances the value of real-time services or improves the efficiency of real-time applications. Location management is a significant burden in IoT due to the resource constraints that need support from edge devices. In the future, make a formal plan to make this study to provide a direction to the IoT domain and to include the suitable mobility models and geographic routing protocols for IoT networks, as it is a key stone in MWNs. Recently, geographical routing protocols are proposed for IoT applications that radically provide new directions in the usage of geographical routing in IoT.

9.4 CONCLUSION

MWNs form a foundation for many modern wireless networks in real-time environments and have increased the necessity of mobility models and geographical routing. The geographical routing is a prominent routing solution to a wide range of MWNs. Realistic mobility models are vital for evaluating the protocols in simulation environments. Mobility models significantly impact various forms of MWNs and its geographical routing solutions. In order to demonstrate the impact of mobility models on various MWNs, this work surveys mobility models and their impact on various geographic routing protocols. Besides the mobility models and their impacts, location management issues and location prediction techniques are discussed. Although several works have been proposed for routing service over various mobility models already, there is a significant scope for future research in geographical routing as well as mobility models. There must be an assurance for better routing performance over realistic mobility models that are suitable for various environments. Finally, this survey concludes with some future directions.

REFERENCES

1. M. Conti, and S. Giordano, "Multi-hop ad-hoc networking: the reality", *IEEE Communications Magazine*, Vol. 45, No. 4, pp. 88–95, 2007.
2. S. Ruhrup, "Theory and practice of geographic routing", *Ad Hoc and Sensor Wireless Networks: Architectures, Algorithms and Protocols*, pp. 1–37, 2009.
3. C. Imrich, M. Conti, and J.J-N. Liu, "Mobile hoc networking: imperatives and challenges", *Ad hoc networks*, Vol. 1, No. 1, pp. 13–64, 2003.
4. I. Bekmezci, O.K. Sahingoz, and Ş. Temel, "Flying ad-hoc networks (FANETs): A survey", *Ad Hoc Networks, Department of Computer Engineering*, Vol. 11, No. 3, pp. 1254–1270, 2013.
5. S. Boussoufa-Lahlaha, F. Semchedinea, and L. Bouallouche-Medjkounea, "Geographic routing protocols for Vehicular Ad hoc NETworks (VANETs): A survey", *Vehicular Communications*, Vol. 11, pp. 20–31, 2018.
6. T. Darwish, K.A. Bakar, and A. Hashim, "Green geographical routing in vehicular ad hoc networks: Advances and challenges", *Computers & Electrical Engineering*, Vol. 64, pp. 436–449, 2017.
7. I. Das, and S. Das, "Geocast routing protocols for ad-hoc networks: comparative analysis and open issues", *Cloud-Based Big Data Analytics in Vehicular Ad-Hoc Networks*, edited by R.S. Rao, et al., IGI Global, 2021, pp. 23–45. Doi: 10.4018/978-1-7998-2764-1.ch002. ISBN. 9781799827641.
8. S. Das, R.S. Raw and I. Das, "Performance analysis of ad hoc routing protocols in city scenario for VANET", *Proceedings of American Institute of Physics, (2nd International Conference on Methods and Models in Science and Technology*, 2011, pp. 257–261, DOI:10.1063/1.3669968.
9. S. Das, R.S. Raw, I. Das and R. Sarkar, "Performance analysis of routing protocols for VANETs with Real Vehicular Traces", *Advances in Intelligent Systems and Computing*, Vol. 243, pp. 45–56, 2014, Doi: 10.1007/978-81-322-1665-0_5.
10. I. Stojmenovic, "Position-based routing in ad hoc networks", *IEEE Communications Magazine*, Vol. 40, No. 7, pp. 128–134, 2002.
11. C. Fraser, K. Curran, J. Santos, and S. Moffett, "A survey of geographical routing in wireless ad-hoc networks", *IEEE Communications Surveys & Tutorials*, Vol. 15, No. 2, pp. 621–653, 2013.
12. K. Nakano, M. Sengoku, and S. Shinoda, "Effect of mobility on connectivity of mobile multi-hop wireless networks", *IEEE 55th Conference on Vehicular Technology*, Vol. 3, pp. 1195–1199, 2002.
13. B. Liu, P. Brass, O. Dousse, P. Nain, and D. Towsley, "Mobility improves coverage of sensor networks", *MobiHoc: Proceedings of the 6th ACM International Symposium on Mobile Ad Hoc Networking and Computing*, pp. 300–308, New York, 2005.
14. M. Grossglauser, and D.N.C Tse, "Mobility increases the capacity of ad hoc wireless networks", *IEEE/ACM Transactions Networks*, Vol. 10, No. 4, pp. 477–486, 2002.
15. S. Capkun, J.-P. Hubaux, and L. Butty´an, "Mobility helps security in ad hoc networks", *MobiHoc: Proceedings of the 4th ACM International Symposium on Mobile Ad Hoc Networking & Computing*, pp. 46–56, Annapolis, MD, 2003.
16. B. Peng, R. Mautz, A. H. Kemp, W. Ochieng and Q. Zeng, "On the effect of localization errors on geographic routing in sensor networks", *Proceedings of the IEEE Communications Society*, pp. 3136–3140, Beijing, China, 2008.
17. D. Son, A. Helmy, and B. Krishnamachari, "The effect of mobility-induced location errors on geographic routing in ad hoc networks: analysis and improvement using mobility prediction", *IEEE Wireless Communications and Networking Conference*, Vol. 1, pp. 189–194, Atlanta, GA, 2004.

18. T. Camp, J. Bowling, and V. Davies, "A survey of mobility models for ad hoc network research", *Wireless Communication and Mobile Computing*, Vol. 2, No. 5, pp. 483–502, 2002.
19. B. Divecha, A. Abraham, C. Grosan, and S. Sanyal, "Impact of node mobility on MANET routing protocols models", *Journal of Digital Information Management*, 4th International Conference on Recent Advances in Information Technology (RAIT), Vol. 5, No. 1, pp. 19–24, 2007.
20. T. Camp, J. Boleng and L. Wilcox, "Location information services in mobile ad hoc networks", *Proceeding IEEE International Conference Communication.*, Vol. 5, pp. 3318–3324, New York, NY, USA, 200.
21. R.S. Raw, D.K Lobiyal, S. Das and S. Kumar, "Analytical evaluation of directional-location aided routing protocol for VANETs", *Wireless Personal Communications*, Springer, Vol. 82, No. 3, pp. 1877–1891, June 2015. Doi: 10.1007/s11277-015-2320-7.
22. J. Treurniet, "A taxonomy and survey of microscopic mobility models from the mobile networking domain", *ACM Computing Surveys (CSUR)*, Vol. 47, No. 1, pp. 1–32, 2014.
23. G. Michal, and K. Grochla, "Review of mobility for performance evaluation of wireless networks", *Advances in Intelligent Systems and Computing*, Springer, Vol. 242, pp. 567–577, 2014.
24. N.P. Vaity, and D.V. Thombre, "A survey on vehicular mobility modeling: flow modeling", *International Journal of Communication Network Security*, Vol. 1, No. 4, p. 21, 2012.
25. R.S. Raw, D.K. Lobiyal and S. Das, "An analytical approach to position-based routing protocol for vehicular ad hoc networks", *Recent Trends in Computer Networks and Distributed Systems Security*, Vol. 335, 2012, pp. 147–152, Doi: 10.1007/978-3-642-34135-9_15.
26. R.S. Raw, D.K Lobiyal, and S. Das, "A probabilistic analysis of border node based MFR routing protocol for vehicular ad-hoc networks", *International Journal of Computer Applications in Technology*, Vol. 51, No. 2, pp. 87–96, 2015.
27. T. He, J.A. Stankovic, C. Lu, and T. Abdelzaher, "SPEED: A stateless protocol for real- time communication in sensor networks", *Distributed Computing Systems, Proceedings. 23rd International Conference IEEE*, pp. 46–55, Providence, RI, 2003.
28. L. Le͂ao and V. Felea, "Latency and lifetime optimization for k-anycast routing algorithm in wireless sensor networks", In N. Montavont, and G. Papadopoulos (eds.), *Ad-hoc, Mobile, and Wireless Networks*. ADHOC-NOW 2018. Lecture Notes in Computer Science, 11104, pp. 39–50. Springer, Cham, 2018. Doi: 10.1007/978-3-030-00247-3_4
29. S. Biswas, and R. Morris, "ExOR: opportunistic multi-hop routing for wireless networks", *ACM SIGCOMM Computer Communication Review*, Vol. 35, No. 4, pp.133–144, 2005.
30. X. Xu, X.-Y. Li, and H. Ma, "Energy-efficient opportunistic routing in wireless sensor networks", *IEEE Transaction on Parallel and Distributed Systems*, Vol. 22, No. 11, pp. 1934–1942, 2011.
31. Y. Li, W. Chen, and Z.-L. Zhang, "Optimal forwarder list selection in opportunistic routing", *Mobile Ad hoc and Sensor Systems. IEEE 6th International Conference*, pp. 670–675, Macau, China, 2009.
32. M. Shelke, A. Malhotra, and P.N. Mahalle, "Congestion-aware opportunistic routing protocol in wireless sensor networks", *Smart Computing and Informatics, Smart Innovation, Systems and Technologies*, Vol. 77, pp. 63–72, Singapore, 2018.
33. S. Souiki, M. Feham, M. Feham, and N. Labraoui, "Geographic routing protocols for underwater wireless sensor networks: a survey", *International Journal of Wireless & Mobile Networks (IJWMN)* Vol. 6, No. 1, pp. 1–19, 2014.

34. K. Ovaliadis and N. Savage, "Cluster protocols in underwater sensor networks: a research review", *Journal of Engineering Science and Technology Review*, Vol. 7, No. 3, pp. 171–175, 2014.
35. S.M. Ghoreyshi, A. Shahrabi, and T. Boutaleb, "Void-handling techniques for routing protocols in underwater sensor networks: survey and challenges", *IEEE Communications Surveys & Tutorials*, Vol. 19, No. 2, pp. 800–827, 2018.
36. X. Fan and F. Du, "An efficient bypassing void routing algorithm for wireless sensor network", *Journal of Sensors*, pp. 1–9, 2015.
37. F. Kuhn, R. Wattenhofer, and A. Zollinger, "An algorithmic approach to geographic routing in ad hoc and sensor networks", *IEEE/ACM Transactions on Networking*, Vol. 16, No. 1, pp. 51–62, 2008.
38. B. Karp and H.T. Kung, "GPSR: greedy perimeter stateless routing for wireless networks", *Proceedings of ACM Mobicom*, pp. 243–254, Boston, MA, 2000.
39. E. Schiller, P. Starzetz, and F. Rousseau, "Binary Waypoint geographical routing in wireless mesh networks", *Proceedings 11th International Symposium on Modeling, Analysis and Simulation of Wireless and Mobile Systems*, p. 252, Vancouver, 2008.
40. J. Na and C. Kim, "GLR: A novel geographic routing scheme for large wireless ad hoc networks", *Computer Networks*, Vol. 50, No. 17, pp. 3434–3448, 2006.
41. J. Na, D. Soroker, and C.-k. Kim, "Greedy geographic routing using dynamic potential field for wireless ad hoc networks", *IEEE Communication Letters*, Vol. 11, No. 3, pp. 243–245, 2007.
42. Y. Koren and J. Borenstein, "Potential field methods and their inherent limitations for mobile robot navigation", *Proceedings 1991 IEEE International Conference on Robotics and Automation*, Vol. 2, No. pp. 1398–1404, Sacramento, CA, 1991.
43. Y.-J. Kim, R. Govindan, B. Karp, and S. Shenker, "On the pitfalls of geographic face routing", *Proceeding Joint Workshop on Foundations of Mobile Computing - DIALM-POMC*, pp. 34–43, Cologne, Germany, 2005.
44. F. Kuhn, R. Wattenhofer, and A. Zollinger, "Asymptotically optimal geometric mobile ad-hoc routing", *Proceeding 6th International Workshop on Discrete Algorithms and Methods for Mobile Computing and Communications - DIALM*, pp. 24–33, Atlanta, GA, 2002.
45. E. Lebhar and Z. Lotker, "Unit disk graph and physical interference model: Putting pieces together", *Parallel and Distributed Processing Symposium, International*, pp. 1–8, 2009.
46. L. Barriere, P. Fraigniaud, and L. Narayanan, "Robust position-based routing in wireless ad hoc networks with unstable transmission ranges", *Proceedings 5th International Workshop on Discrete Algorithms and Methods for Mobile Computing and Communications ACM*, pp. 19–27, 2001.
47. F. Kuhn, R. Wattenhoffer, and A. Zollinger, "Worst-case optimal and average-case efficient geometricad-hocrouting", *Proceedings 4th ACM International Symposium on Mobile Ad Hoc Networking & Computing, ser. MobiHoc*, ACM, pp. 267–278, Annapolis, MD, 2003.
48. F. Kuhn, R. Wattenhofer, Y. Zhang, and A. Zollinger, "Geometric ad-hoc routing: of theory and practice", *Proceedings of the 22nd ACM International Symposium on the Principles of Distributed Computing(PODC)*, pp. 63–72, Boston, MA, 2003.
49. B. Leong, S. Mitra, and B. Liskov, "Path vector face routing: geographic routing with local face information", *13th IEEE International Conference on Routing Protocols*, p. 12, Boston, MA, 2005.
50. H. Frey and I. Stojmenovic, "On delivery guarantees of face and combined greedy-face routing in ad hoc and sensor networks", *Proceedings of the 12th Annual International Conference on Mobile Computing and Networking*, pp. 390–401, Los Angeles, CA, 2006.

51. S. Biswas, and R. Morris, "Opportunistic routing in multi-hop wireless networks", *ACM SIGCOMM Computer Communication Review*, Vol. 34, No. 1, pp. 69–74, 2004.
52. S. Chachulski, M. Jennings, S. Katti, and D. Katabi, "Trading structure for randomness in wireless opportunistic routing", *Proceedings of the Conference on Applications, Technologies, Architectures, and Protocols for Computer Communications*, ACM, Vol. 37, No. 4, pp. 169–180, Kyoto, Japan, 2007.
53. Z. Zhong, and S. Nelakuditi, "On the efficacy of opportunistic routing", *4th Annual IEEE Communications Society Conference on Sensor, Mesh and Ad Hoc Communications and Networks*, Vol. 7, pp. 441–450, San Diego, CA, 2007.
54. Z. Zhong, J. Wang, S. Nelakuditi, and G.H. Lu, "On selection of candidates for opportunistic any path forwarding", *ACM SIGMOBILE Mobile Computing and Communications Review*, Vol. 10, No. 4, pp. 1–2, 2006.
55. M. Zorzi, & R.R. Rao, "Geographic random forwarding (GeRaF) for ad hoc and sensor networks: Multi-hop performance", *IEEE Transactions on Mobile Computing*, Vol. 2, No. 4, pp. 337–348, 2003.
56. I. Stojmenovic and X. Lin, "Loop-free hybrid single-path/flooding routing algorithms with guaranteed delivery for wireless networks", *IEEE Transactions on Parallel and Distributed Systems*, Vol. 12, No. 10, pp. 1023–1032, 2001.
57. R. Jain, A. Puri, and R. Sengupta, "Geographical routing using partial information for wireless ad hoc networks", *IEEE Personal Communications*, Vol. 8, No. 1, pp. 48–57, 2001.
58. D. chen and P.k. Varshney, "A survey of void handling techniques for geographic routing in wireless networks", *IEEE Communications Surveys and Tutorials*, Vol. 9, No. 1, pp. 50–67, 2007.
59. B. Leong, B. Liskov, and R. Morris, "Geographic routing without planarization", *Proceeding 3rd conference on Networked Systems Design & Implementation, USENIX Association*, Vol. 3, pp. 25–25, Berkeley, CA, 2006,
60. S. Ruhrup and I. Stojmenovic, "Contention-based georouting with guaranteed delivery, minimal communication overhead, and shorter paths in wireless sensor networks", *Parallel Distributed Processing (IPDPS), IEEE International Symposium*, pp. 1–9, Atlanta, GA, 2010.
61. H. Kalosha, A. Nayak, S. Ruhrup, and I. Stojmenovic, "Select-and-protest-based beaconless georouting with guaranteed delivery in wireless sensor networks", *27th IEEE Conference on Computer Communications*, pp. 346–350, Phoenix, AZ, 2008.
62. S. Ruhrup, H. Kalosha, A. Nayak, and I. Stojmenovic, "Message- Efficient beaconless georouting with guaranteed delivery in wireless sensor, AdHoc, and actuator networks", *IEEE/ACM Transactions on Networking*, Vol. 18, No. 1, pp. 95–108, 2010.
63. F. Zhao and L. J. Guibas, *Wireless Sensor Networks: An Information Processing Approach* Morgan Kaufmann Publishers, 2004.
64. A. Omer, and K. Ahmed, "Analytical study of MFR routing algorithm for mobile ad hoc networks", *Journal of King Saud University-Computer and Information Sciences*, Vol. 22, pp. 29–35, 2010.
65. E. Kranakis, H. Singh, and J. Urrutia, "Compass routing on geometric networks", *Proceeding of 11th Canadian Conference on Computational Geometry*, Canada, 1999.
66. I. Stojmenovic and X. Lin, "Power-aware localized routing in wireless networks", *IEEE Transactions on Parallel and Distribution System*, Vol. 12, No. 11, pp. 1122–1133, 2001.
67. H. Hwang, I. Hur, and H. Choo, "GOAFR plus-ABC: geographic routing based on adaptive boundary circle in MANETs", *International Conference on Information Networking*, Chiang Mai, India, 2009.
68. B. Newton, J. Aikat, and K. Jeffay, "Geographic routing in large-scale highly- dynamic mobile ad hoc networks", *IEEE 24th International Symposium on Modeling, Analysis and Simulation of Computer and Telecommunication Systems(MASCOTS)*, London, UK, 2015.

69. S. Yang, F. Zhong, C.K. Yeo, B.S. Lee, and J. Boleng, "Position based opportunistic routing for robust data delivery in MANETs", *Global Telecommunications Conference, GLOBECOM*. IEEE, pp. 1–6, Honolulu, HI, USA, 2009.
70. A.S. Cacciapuoti, M. Caleffi, and L. Paura, "A theoretical model for opportunistic routing in ad hoc networks", *International Conference on Ultra Modern Telecommunications & Workshops*, IEEE, St. Petersburg, Russia, 2009.
71. C.-L. Hu and C. Sosorburam, "Enhanced geographic routing with two-hop neighborhood information in sparse MANETs", *Wireless Personal Communications: An International Journal Archive*, Vol. 107, No. 1, pp. 417–436, 2019.
72. M. Rieke, T. Foerster, A. Broering, "Unmanned aerial vehicles as mobile multi-platforms", *The 14th AGILE International Conference on Geographic Information Science*, pp. 18–21, 2011.
73. Z. Han, A.L. Swindlehurst, K.J.R. Liu, "Optimization of MANET connectivity via smart deployment/movement of unmanned air vehicle", *IEEE Transactions on Vehicular Technology*, Vol. 58, No. 7, pp. 533–3546, 2009.
74. S. Lee, B. Bhattacharjee, S. Banerjee and Bo Han, "A General Framework for Efficient Geographic Routing in Wireless Networks", *Computer Networks*, Vol. 54, No. 5, pp. 844–861, 2010.
75. B.A. Mahmood and D. Manivannan, "GRB: greedy routing protocol with backtracking for mobile ad hoc networks", *IEEE 12th International Conference on Mobile Ad Hoc and Sensor Systems*, Dallas, TX, 2015.
76. D. Yang, H. Xia, E. Xu, D. Jing and H. Zhang, "An energy- balanced geographic routing algorithm for mobile ad hoc networks", *Energies*, Vol. 11, No. 9, pp. 1–16, 2018.
77. S. Hyeon, K.-I. Kim, S. Yang, "A new geographic routing protocol for aircraft ad hoc networks", *Proceedings of the 29th IEEE Digital Avionics Systems Conference (DASC)*, Salt Lake City, UT, 2010.
78. E. Kuiper, S. Nadjm-Tehrani, "Geographical routing with location service in intermittently connected MANETs", *IEEE Transactions on Vehicular Technology*, Vol. 60, No. 2, pp. 592–604, 2011.
79. A. Guillen-Perez, and M.D. Cano, "Flying ad hoc networks: A new domain for network communications", *Sensors*, Vol. 18, No. 10, p. 3571, 2018.
80. J.P. Jabbar "AeroRP: A geolocation assisted aeronautical routing protocol for highly dynamic telemetry environments", *Proceedings of the International Telemetering Conference*, pp. 1–10, Las Vegas, NV, 2009.
81. D. Rosario, Z. Zhao, T. Braun, E. Cerqueira, A. Santos, and I. Alyafawi, "Opportunistic routing for multi-flow video dissemination over flying ad-hoc networks", *Proceedings of the 15th International IEEE Symposium on a World of Wireless, Mobile and MultimediaNetworks (WoWMoM)*, pp. 1–6, Sydney, NSW, 2014.
82. J. Zhang, T. Chen, S. Zhong, J. Wang, W. Zhang, X. Zuo, R.G. Maunder, L.H. Fellow, "Aeronautical ad-hoc networking for the internet-above- the-clouds", *Proceedings of the IEEE*, Vol. 107, No. 5, pp. 868–911, 2019.
83. S. Wang, C. Fan, C. Deng, W. Gu, Q. Sun, F. Yang, "A- GR: a novel geographical routing protocol for AANETs", *Journal of Systems Architecture*, Vol. 59, No. 10, pp. 931–937. 2013.
84. M.Z. Zamalloa, K. Seada, B. Krishnamachari, and A. Helmy, "Efficient geographic routing over lossy links in wireless sensor networks", *ACM Transactions on Sensor Networks (TOSN)*, Vol. 4, No. 3, pp. 1–33, 2008.
85. K. Seada, M. Zuniga, A. Helmy, and B.K. machari, "Energy- efficient forwarding strategies for geographic routing in lossy wireless sensor networks", *ACM Proceedings of the 2nd International Conference on Embedded Networked Sensor Systems*, pp. 108–121, Baltimore, MD, 2004.
86. J.N. Al-Karaki, and A.E. Kamal, "Routing techniques in wireless sensor networks: a survey", *IEEE Wireless Communications*, Vol. 11, No. 6, pp. 6–28, 2004.

87. M. I. Khan, W.N. Gansterer, and G. Haring. "Static vs. mobile sink: the influence of basic parameters on energy efficiency in wireless sensor networks", *Computer Communications*, Vol. 36, No. 9, pp. 965–978, 2013.
88. K. Akkaya, and M. Younis, "A survey on routing protocols for wireless sensor networks", *Ad hoc Networks*, Vol. 3, No. 3, pp. 325–349, 2005.
89. T.P. Lambrou, and C.G. Panayiotou, "A survey on routing techniques supporting mobility in sensor networks", *5th International Conference on Mobile Ad-Hoc and Sensor Networks*, pp. 78–85, Fujian, China, 2009.
90. C. Tunca, S. Isik, M.Y. Donmez, and C. Ersoy, "Distributed mobile sink routing for wireless sensor networks: A survey", *IEEE Communications Surveys & Tutorials*, Vol. 16, No. 2, pp. 877–897, 2014.
91. D. Yang, H. Xia, E. Xu, D. Jing, and H. Zhang, "An energy-balanced geographic routing algorithm for mobile ad hoc networks", *Energies*, Vol. 11, No. 9, p. 2219, 2018.
92. M. Khelifi, S. Bourouais, O. Lounis, and S. Moussaoui, "GRCS: A cluster-based geographic routing protocol for WSNs", *Ninth International Conference on Ubiquitous and Future Networks (ICUFN) IEEE*, pp. 249–254, Milan, Italy, 2017.
93. M. Tommaso, P. Dario, C. Vehbi, and I. F. Akyildiz, "A distributed coordination framework for wireless sensor and actor networks", *Proceedings of the 6th ACM International Conference*, Urbana-Champaign, Illinois, USA, pp. 99–1, 2005.
94. L. Yen-Ting and S. Megerian, "Low cost distributed actuation in large-scale ad hoc sensor–actuator networks", *International Conference on Wireless Networks, Communications and Mobile Computing*, Vol. 2, pp. 975–980, Maui, HI, 2005.
95. M. Rekik, N. Mitton, and Z. Chtourou, "GRACO: a geographic GReedy routing with an ACO-based void handling technique", *International Journal of Sensor Networks*, Vol. 26, No. 3, pp. 145–161, 2018.
96. Joshi, G.P., and S.W. Kim, "A distributed geo-routing algorithm for wireless sensor networks", *Sensors*, Vol. 9, No. 6, pp. 4083–4103, 2009.
97. E.-S. Cho, Y. Yim, and S.-H. Kim, "Transfer-efficient face routing using the planar graphs of neighbors in high density WSNs", *Sensors*, Vol. 17, No. 10, pp. 1–20, 2017.
98. P. Xie, J. Cui, and L. Lao, "VBF: Vector-based forwarding protocol for underwater sensor networks", *IFIP International Federation for Information Processing*, pp. 1216–1221, 2006.
99. N. Nicolaou, A. See, P. Xie, J.H. Cui, and D. Maggiorini, "Improving the robustness of location-based routing for underwater sensor networks", *IEEE Oceans Conference*, pp. 1–6, Aberdeen, 2007.
100. C. Jinming, W. Xiaobing, and C. Guihai, "REBAR: a reliable and energy balanced routing algorithm for UWSNs", *Proceedings of the Seventh International Conference on Grid and Cooperative Computing*, pp. 349–355, Shenzhen, China, 2008.
101. J.M. Jornet, M. Stojanovic, and M. Zorzi, "Focused beam routing protocol for underwater acoustic networks", *Proceeding of the third ACM International Workshop on UnderWater Networks WUWNet*, pp. 75–82, San Francisco, CA, 2008.
102. H. Daeyoup, and K. Dongkyun, "DFR: directional flooding-based routing protocol for underwater sensor networks" *IEEE OCEANS*, pp. 1–7, Quebec City, 2008.
103. N. Chirdchoo, S. Wee-Seng, and C. Kee Chaing, "Sector-based routing with destination location prediction for underwater mobile networks", *Proceedings of the International Conference on Advanced Information Networking and Applications Workshops*, pp. 1148–1153, Bradford, 2009.
104. KR. Anupama, A. Sasidharan, and S. Vadlamani, "A location-based clustering algorithm for data gathering in 3D underwater wireless sensor networks", *Proceedings of the International Symposium on Telecommunications, IST*, pp. 343–348, Tehran, Iran, 2008.
105. S.M. Ghoreyshi, A. Shahrabi, and T. Boutaleb, "A stateless opportunistic routing protocol for underwater sensor networks", *Wireless Communications and Mobile Computing*, Vol. 2018, pp. 1–18, 2018.

106. Z. Rahman, F. Hashim, M.F.A. Rasid, and M. Othman, "Totally opportunistic routing algorithm (TORA) for underwater wireless sensor network", *PLoS One*, Vol. 13, No. 6, pp. 1–28, 2018.
107. V. Naumov, R. Baumann, and T. Gross, "An evaluation of inter-vehicle ad hoc networks based on realistic vehicular traces", *Proceedings of the 7th ACM International Symposium on Mobile Ad Hoc Networking and Computing*, pp. 108–119, Florence, Italy, 2006.
108. C. Lochert, H. Hartenstein, J. Tian, H. Fussler, D. Hermann, and M. Mauve, "A routing strategy for vehicular ad hoc networks in city environments", *IEEE Intelligent Vehicles Symposium*, pp. 156–161, Columbus, OH, 2003.
109. H. Fuler, H. Hartenstein, M. Mauve, W. Effelsberg, and J. Widmer, "Contention-based forwarding for street scenarios", *1st International Workshop in Intelligent Transportation*, WIT, 2004.
110. R.S. Raw and S. Das, "Performance analysis of P-GEDIR protocol for vehicular ad hoc network in urban traffic environments", *Wireless Personal Communication*, Springer, Vol. 68, No. 1, pp. 65–78, 2013.
111. R.S. Raw, S. Das, N. Singh, and S. Kumar, "Feasibility evaluation of VANET using Directional-Location Aided Routing (D-LAR) protocol", *International Journal of Computer Science Issues*, Vol. 9, No. 5, pp. 404–410, 2012.
112. D. Borsetti, and J. Gozalvez, "Infrastructure-assisted geo-routing for cooperative vehicular networks", *IEEE Vehicular Networking Conference (VNC)*, pp. 255–262, Jersey City, NJ, 2010.
113. S. Tsiachris, G. Koltsidas, and F.-N. Pavlidou, "Junction-based geographic routing algorithm for vehicular ad hoc networks", *Wireless Personal Communications*, Vol. 71, No. 2, pp. 955–973, 2013.
114. H. Kang and D. Kim, "Vector routing for delay tolerant networks", *Proceedings of the 68th IEEE Vehicular Technology Conference (VTC-Fall)*, pp. 1–5, Calgary, Canada, 2008.
115. O. Turkes, H. Scholten, and P. Havinga, "RoRo-LT: social routing with next-place prediction from self-assessment of spatiotemporal routines", *Proceedings of the 10th International Conference on Ubiquitous Intelligence and Computing and 10th International Conference on IEEE Autonomic and Trusted Computing (UIC/ATC)*, pp. 201–208, Vietri sulMare, Italy, 2013.
116. R. Karimi, and S. Shokrollahi, "PGRP: predictive geographic routing protocol for VANETs", *Computer Networks*, Vol. 141, pp. 67–81, 2018.
117. J. LeBrun, C.N. Chuah, D. Ghosal, and M. Zhang. "Knowledge-based opportunistic forwarding in vehicular wireless ad hoc networks", *IEEE Proceedings of the 2005 IEEE Vehicular Technology Conference*, pp. 2289–2293, Stockholm, Sweden, 2005.
118. D. Huang, Y. Yan, C. Su, and G. Xu, "Prediction-based geographic routing over VANETs", *Rev. Téc. Ing. Univ. Zulia*, Vol. 39, No. 2, pp. 157–164, 2016.
119. I. Leontiadis and C. Mascolo, "Geopps: Geographical opportunistic routing for vehicular networks", *Proceedings of the IEEE International Symposium on World of Wireless, Mobile and Multimedia Networks*, pp. 1–6, Espoo, Finland, Finland, 2007.
120. S. Boussoufa-Lahlah, F. Semchedine, and L. Bouallouche-Medjkoune, "Geographic routing protocols for Vehicular Ad hoc NETworks (VANETs): A survey", *Vehicular Communications*, Vol. 11, pp. 20–31, 2018.
121. L. Lemus Cárdenas, A. Mohamad Mezher, and N.P. López Márquez, "3MRP+: an improved multimetric geographical routing protocol for VANETs", *Proceedings of the15th ACM International Symposium on Performance Evaluation of Wireless AdHoc, Sensor, & Ubiquitous Networks*, pp. 33–39, Montreal, 2018.
122. N. Li, J.-F. Martinez-Ortega, V.H. Diaz, J.A.S. Fernandez, "Probability Prediction based Reliable Opportunistic (PRO) routing algorithm for VANETs", *IEEE/ACM Transactions on Networking*, Vol. 26, No. 4, pp. 1–13, 2018.

123. I. Leontiadis, P. Costa, and C. Mascolo, "Extending access point connectivity through opportunistic routing in vehicular networks", *Proceedings IEEE INFOCOM*, San Diego, CA, USA, 2010.
124. S. Kim, "Timed bargaining-based opportunistic routing model for dynamic vehicular ad hoc network", *Journal on Wireless Communications and Networking*, Vol. 1, pp. 1–9, 2016.
125. C. Chen, L. Liu, T. Qiu, K. Yang, F. Gong, and H. Song, "ASGR: an artificial spider-web-based geographic routing in heterogeneous vehicular networks", *IEEE Transactions On Intelligent Transportation Systems*, Vol. 20, No. 5, pp. 1604–1620, 2018.
126. A.N. Hassan, A.H. Abdullah, O. Kaiwartya, Y. Cao, and D.K. Sheet, "Multi-metric geographic routing for vehicular ad hoc networks", *Wireless Networks*, Vol. 24, No. 7, pp. 2763–2779, 2018.
127. K. Aravindhan, C.S.G. Dhas, "Destination-aware context-based routing protocol with hybrid soft computing cluster algorithm for VANET", *Ad Hoc Networks*, Vol. 9, No. 2, pp. 131–141, 2018.
128. R.S. Bali, N. Kumar, and J.J. Rodrigues, "Clustering in vehicular ad hoc networks: Taxonomy, challenges and solutions", *Vehicular Communications*, Vol. 1, No. 3, pp. 134–152, 2014.
129. M. Ren, L. Khoukhi, H. Labiod, J. Zhang, and V. Vèque, "A mobility-based scheme for dynamic clustering in vehicular ad-ho cnetworks(VANETs)", *Vehicular Communications*, Vol. 9, pp. 233–241, 2017.
130. C. Caini, H. Cruickshank, S. Farrell, and M.Marchese, "Delayand disruption-tolerant networking (DTN): an alternative solution for future satellite networking applications", *Proceedings of the IEEE*, vol. 99, No. 11, pp. 1980–1997, 2011.
131. T. Wang, Y. Cao, Y. Zhou, and P. Li, "A survey on geographic routing protocols in delay/disruption tolerant networks", *International Journal of Distributed Sensor Networks*, Vol. 12, No. 2, 3174670, 2016.
132. J. LeBrun, C.-N. Chuah, D. Ghosal, and M. Zhang, "Knowledge based opportunistic forwarding in vehicular wireless Ad Hoc networks", *Proceedings of the IEEE 61st Vehicular Technology, Conference (VTC)*, pp. 2289–2293, Stockholm, Sweden, 2005.
133. I. Leontiadis and C. Mascolo, "GeOpps: geographical opportunistic routing for vehicular networks", *Proceedings of the IEEE International Symposium on a World of Wireless, Mobile and Multimedia Networks (WOWMOM)*, pp. 1–6, Espoo, Finland, 2007.
134. V.N.G.J. Soares, J.J.P C. Rodrigues, and F. Farahmand, "GeoSpray: a geographic routing protocol for vehicular delay tolerant networks", *Information Fusion*, Vol. 15, No. 1, pp. 102–113, 2014.
135. K. Peters, A. Jabbar, E. K. Cetinkaya, and J.P.G. Sterbenz, "A geographical routing protocol for highly-dynamic aeronautical networks", *Proceedings of the IEEE Wireless Communications and Networking Conference (WCNC)*, pp. 492–497, Cancun, Mexico, 2011.
136. Y. Cao, Z. Sun, N. Wang, H. Cruickshank, and N. Ahmad, "A reliable and efficient geographic routing scheme for delay/disruption tolerant networks", *IEEE Wireless Communications Letters*, Vol. 2, No. 6, pp. 603–606, 2013.
137. H.-Y. Huang, P.-E. Luo, M. Li et al., "Performance evaluation of SUVnet with real-time traffic data", *IEEE Transactions on Vehicular Technology*, Vol. 56, No. 6, pp. 3381–3396, 2007.
138. X. Li, W. Shu, M. Li, H. Huang, and M.-Y. Wu, "DTN routing in vehicular sensor networks", *Proceedings of the IEEE Global Telecommunications Conference (GLOBECOM)*, pp. 752–756, New Orleans, LA, 2008.
139. E. Kuiper and S. Nadjm-Tehrani, "Geographical routing with location service in intermittently connected MANETs", IEEE Transactions on Vehicular Technology, Vol. 60, No. 2, pp. 592–604, 2011.

140. J.A.B. Link, D. Schmitz, and K. Wehrle, "GeoDTN: geographic routing in disruption tolerant networks", *Proceedings of the 54th Annual IEEE Global Telecommunications Conference (GLOBECOM)*, pp. 1–5, Houston, TX, 2011.
141. S. Kumar and R.S. Raw, "Minimize the Routing Overhead through 3D Cone Shaped Location-Aided Routing Protocol for FANETs", *International Journal of Information Technology*, Vol. 13, No. 1, pp. 89–95, 2020.
142. S. Kumar, A. Bansal, and R. S. Raw, "Analysis of Effective Routing Protocols for Flying Ad-Hoc Networks", *International Journal of Smart Vehicles and Smart Transportation*, Vol. 3 No. 2, pp. 1–18, 2020, Doi: 10.4018/IJSVST.2020070101.
143. P. Singh, R. S. Raw, S. A. Khan, M. A. Mohammed, A. A. Aly and D.-N. Le, "W-GeoR: weighted geographical routing for VANET's health monitoring applications in urban traffic networks", *IEEE Access*, Doi: 10.1109/ACCESS.2021.3092426.
144. P. Singh, R. S. Raw and S. A. Khan, "Link risk degree aided routing protocol based on weight gradient for health monitoring applications in vehicular ad-hoc networks", *Journal of Ambient Intelligence Humanized Computing*, 2021, Doi: 10.1007/s12652-021-03264-z.
145. P. Sing and R.S. Raw "Development of novel framework for patient health monitoring system using VANET: an Indian perspective", *International Journal of Information Technology*, pp. 251–260, 2020.

10 Energy-Aware Secure Routing in Sensor Network

N. Ambika
St.Francis College, Bangalore, India

CONTENTS

10.1 Introduction 173
10.2 Literature Survey 174
10.3 Assumptions Considered in the Proposal 183
10.4 Proposed Work 183
10.5 Simulation 184
    10.5.1 Energy Consumption 184
10.6 Conclusion 185
References 185

10.1 INTRODUCTION

The trend nowadays is demanding automation to minimize human efforts and increase productivity. It is not only in the industry-related sets but also in routine domains. Sensors (Akyildiz, Su, Sankarasubramaniam, & Cayirci, 2002; Ambika, 2020a) have given comfort and are catering to the needs to a large extent despite their setbacks. The enormous devices deploy to gather readings from the environment. These tiny devices (Ambika, 2020b) are deployed in many applications to monitor the surroundings, process the information, and transmit it to the destination. The devices are small that fit in the palm, and this property keeps them from being unnoticed. They find the usage in military surveillance (Lee, Hyuk, Lee, Song, & Lee, 2009), monitoring heavy traffic (Deng, Han, & Mishra, 2006), contamination examination (Khedo, Perseedoss, & Mungur, 2010), detect forest blaze (Albert, et al., 2008), water quality analysis (Baoding & Liu, 2010; Peijina, Tingb, & Yandongc, 2011), home monitoring (El-Basioni, El-Kader, & Abdelmonim, 2013), wearables providing continuous readings (Chen, Lee, Chen, Huang, & Luo, 2009), disaster management (Ahmed, Bakar, Channa, Khan, & Haseeb, 2017), etc. These instruments have some drawbacks. They lack energy, bandwidth, and memory parameters necessary to keep the network in a good working state for a long time.

SDARP (Kumar, Jayasankar, Eswaramoorthy, & Nivedhitha, 2020) considers two kinds of nodes in its study: nonmobile and mobile devices (Hu & Evans, 2004).

DOI: 10.1201/9781003206453-10

Mobile devices supervise the traffic. The nonmobile nodes record the readings and provide the same to the destination. The grouping (Xiangning, 2007) uses the count of nodes and proximity from one to another. A voting algorithm finds the group head among the cluster members. Other parameters considered in selection include device capability, dynamic nature, and steadiness, and transmission intensity. The group head supervises. The group head indicates to gather data. In the route discovery procedure, the group head looks for the lookup table and assembles the data. The time delay is determined. This methodology hinders excessive packet assembly. If the group head has any fluctuation, it notifies the Optimal Cluster head. The system adopts a power-consuming encoding methodology. It generates a secret and shared credential. The cipher uses energy and a confidential key. The head receives the decrypted score and modulus score metric.

The proposal tries to improve the previous contribution by minimizing energy consumption. This methodology aims to increase the lifespan of the nodes in the network. The suggestion consists of three types of nodes in the network—static nodes deployed to sense the surroundings, assisting nodes that provide traffic of the environment, and mobile nodes that forward the packets to the next available hop. In this work, the auxiliary nodes analyze the traffic and update the lookup table. This method minimizes the load on the static nodes and increases their lifespan. The proposal also adopts a dual methodology to choose the head of the cluster. The choice is made based on the assaults in the environment. The work decreases energy consumption by 4.038% compared with the previous contribution.

The work is divided into seven divisions. The motive of the work is narrated in the motivation section. A detailing of the previous contribution is briefed in Section 10.3. Some assumptions are made in the work. The same is detailed in Section 10.4. The work is detailed in Section 10.5. The simulation details are narrated in segment six. The work is concluded in Section 10.7.

10.2 LITERATURE SURVEY

In Parno, Luk, Gaustad, and Perrig (2006) allows the web address to all the devices. It creates the course-plotting entries by adopting a clustering procedure. The program prevents assaults based on the information of the entries. It limits the disrupted devices to connect to the other nodes of the network. All the instruments of the system own the system address and the course-plotting entries. It links the address to the next-door device. The constructed tree routes the packets using the course-plotting. It binds the storage on the course-plotting, assigns addresses, and counts the route length. The system is capable of detecting a device. The evaluation tree verifies the divergence and executes the clustering algorithm. The Honeybee methodology enforces identifying the guilty. This procedure detaches the illegitimate device from the network. The packets are routed from one end of the network to another record in the course-plotting storage. The source has a threshold over the number of the path chosen to transmit its data. Energy is a limitation. The device willing sends data collects information of the next-door device of its transmission. If the transmission size is less than the threshold, it opts to merge the transmitting packets and dispatch them to the destination. In this scenario, the devices installed in the borders connect to the clusters nearby. They create merge outcomes and transmit them to the group.

The system (Du, Guizani, Xiao, & Chen, 2007) has two types of nodes. Some of the nodes are better in capability than the other category. These nodes arbitrarily deploy in the environment. The nodes communicate with each other to form groups. The more capable nodes serve as the chief. The controller assembles the data from various devices of the group. They eliminate the replication and forward to the next available hop. The credential is created by exchanging information between the two parties. They undergo duel handshakes using the message authentication codes. A least parsing tree constructs to route the packets. If the chiefs do not merge the data, it uses a quick route to the destination.

The work (Du, Xiao, Chen, & Wu, 2006) suggests a safe transmission methodology. The trio-handshake procedure finds its usage in the method. The nodes protect against Hello overflow assaults. The devices create a group based on their locale. It calculates the time required to locate an instrument and agrees to negotiate. They embed the common credential between the device and the sink. It uses an imitated arbitrary method. It stores in the sink. The host computes the device identification using the device's unique number.

The devices (Qin, Yang, Jia, Zhang, Ma, & Ding, 2017) monitor other nodes and detect the illegitimate node. It computes the reliability of the devices in their surroundings. The logical tree procedure is adopted to trace a safe transmission of the packets. Direct character analysis is measured. They are in observation by other devices of the network. Based on the decomposing period feature, it undergoes an evaluation of the count of assessing files. It considers the reliability ratio to transmit their data. The length of packets, the proximity between the devices, and power spent are some of the parameters used to maintain a balance in the network. It also exchanges the trust values among themselves in the second method. The received data from the neighbor is verified, and divergent values are detected. The power consumption consideration is in this methodology. It uses the nonparticipation ratio of the devices and the overindulgence behavior of the nodes.

The proposed work (Madria & Yin, 2009) minimizes wormhole assaults in the environment. The protocol falls into four phases. To know each other in the network, the devices ping messages in the neighborhood. The receiver acknowledges the data. The next-door device lists the acknowledgments. The instrument also exchanges the record with other nodes and makes a comparison with its catalog. If the register entry matches with each other, they conclude the nonexistence of the assaults. The two communicating parties create a credential using an arbitrary function and key. The sink transmits the path to follow to reach the destination. The devices receiving the message follow the same instructions. A new track is created by the nodes using the next-door register entries. The sink makes and broadcasts a new path if it suspects the packet count it has received from the network.

The authors (Yang, Xiangyang, Peng, Tonghui, & Leina, 2018) have used three parameters in their proposal. The proposal uses a power-conserving protected transmission methodology. The procedure tracks to evaluate trust, construction of the route, and its conservation. The reliability between the devices is estimated based on their performance during the transmission of data. The path creates using the reliability estimate. The estimation also uses the number of leaps the data has to take to reach the destination. The deployed devices calculate the secondary reliability value. The system undergoes verification of the reliability. They are on constant vigilance analyzing the transmission conduct. The messages evaluate in case of a decline in count and meddling of them. Beta dispersal examines neighbor activities. An atypical reduction feature incorporates avoiding external factors like losing the information due to a bottleneck. The devices also collect the data from the next-door nodes. The messages introduce a tolerance level. It calculates the reliability quantity using the registry entries of it and the next-door nodes.

It listens to their communication conduct to evaluate the devices for their legitimacy (Ahmed, Bakar, Channa, & Khan, 2016). They detect illegitimate devices using the procedure. The entries regarding the transmission are maintained to compare with the toleration level. The instrument categorizes into three classes: legitimate, guilty, and defective set. The beta possibility denseness procedure is used in gamma functionality to conclude the conduct of the devices. The work contains a graphical illustration consisting of connections between the devices. The reliable devices are taken into consideration to make a consistent track for transmission. The amalgamated steering utility measurement chooses an optimal path. The work also suggests some methodologies to solve the path issue. If the devices are short with energy, they notify their parents. On spotting any malicious activity, the devices generate a report regarding the same. The host receives the narration of transmission for suggesting an alternate path.

The work presented (Ahmed, Bakar, Channa, Khan, & Haseeb, 2015) considers creating a reliable model. It aids in path identification and preservation. The reliability model is based on one-one surveillance, third-party monitoring, and generated affirmed conduct of the devices. The necessary information to draw out a conclusion is maintained. It includes the device identity, one-one reliability report, third-party reliability report, anticipated affirmation conduct report, summated conduct report, and reported time interval. The model detects an optimal path considering the power utilization. The beta dispersal calculation evaluates the device for its legitimacy.

The model (Fang, Zhang, Chen, Liu, & Tang, 2019) improves the reliability and security of the environment. The reliability record creation has the previous doings of the device. The model checks the device for Gaussian dispersal. The procedure uses a collection of reciprocal data and computes variance. The old value and subsequent probable denseness are estimated. One-one reliability uses interrelate activities count for prior doings and Gaussian dispersal estimate. The epoch period jolts all the previous doings. Using the gray hypothesis, collection of factors is concluded. The reliability measure and the jump count are arrived at a decision. The properties of the device are evaluated and assessed. The price is an entry in the matrix.

The system applies normalization to the parts of the matrix. The work considers the chosen idyllic element. The estimation grades the position. Pareto credential aids in optimization (Sun, Wei, Zhang, & Qu, 2019). Secure Routing Protocol based on Multi-objective Ant-colony-optimization (SRPMA) improves the old-style single target insect and enhances directing calculation to get ideal steering by thinking about the lingering energy objective and the trust esteem objective.

It introduces swarming distance. The arrangements are in rising request and set limit. It is for every gadget. The work rules out the same. It builds two sorts of heuristic data inferring the two destinations. The state progress rules the following position directing hub dependent on the maximum state change likelihood. The state progress likelihood is a proportion of insect of a gadget to pick gateway another as the following bounce hub at the hub. The neighborhood pheromone refreshing guideline is that the pheromone of every target of each arrangement way. It promptly refreshes when every subterranean insect has developed a doable arrangement by utilizing the recipe with an alternate augmentation equation. The pheromone is relative to the lingering energy and trust estimation of the steering way. The nodes with more trust and more leftover energy will have more opportunities as steering hubs. Having finished the neighborhood pheromone restoring and it forwards ants in reverse. The sink hub will refresh the external file. It includes the non-ruled arrangements with the improved swarming distance model. The worldwide pheromone refreshing standard is that the entirety of the refreshed non-ruled arrange in the outer chronicle will be restored by the retrogressive ants as indicated by the equation utilizing the worldwide pheromone increase equation.

The authors (Lu, Li, & Kameda, 2010) consider a Wireless Sensor Network (WSN) system. It consists of a fixed base station and a large number of sensor nodes. The security uses bilinear Diffie-Hellman problem (BDHP) in the pairing domain. The proposal is an ID-based secure routing protocol. Time divides into successive time intervals. It uses time-stamps. The proposed secure routing protocol operates in rounds based on Low Energy Adaptive Clustering Hierarchy (LEACH). Each step has a setup phase and a steady-state phase. With the synchronization of time, sensor nodes know when each stage starts and ends. It addresses the solution management of ID-based cryptography for secure routing. It introduces the operation of the routing protocol. Upon receiving the message, each sensor node verifies its authenticity. It checks the time-stamp of the current time interval and determines whether the received message is fresh. Those who decide to become Cluster Heads (CHs) have to broadcast an advertisement message to all sensor nodes in the network. It concatenates with the digital signature. The setup phase consists of four steps. It broadcasts its identification, the start time of the current round, and a nonce to all sensor nodes. Then the sensor nodes decide whether to become CH for the current step based on the threshold and compare them with the predetermined random number from 0 to 1. Once the setup phase is over, the network system turns into the steady-state phase, in which sensed data transmits from sensor nodes to the Base Station (BS). The steady-state phase consists of multiple reporting cycles of data transmissions from leaf nodes to the CHs and is exceedingly long compared to the setup phase.

The gathering chief (Yin & Madria, 2006) infers its mystery key. Second, it verifies its neighboring gathering pioneer mystery key. Third, it sets up the between bunch imparted solution to its neighboring gathering pioneers. It processes the pairwise shared key. Fourth, after a predetermined time, all sensor hubs other than the sink hub eliminate every mystery key. The gathering chief sends the message to its one-bounce neighbors. The neighbors at that point confirm the verification of the message through MAC. At that point, the neighbors set up the intra-bunch shared key.

The gathering chief eliminates the underlying answer, the irregular capacity, and the arbitrary personality. The information affirmation system identifies the dark opening assault because of the agreement between the gathering chief and different hubs. It is dependent on the perception that the dark opening hub will drop the information parcels during the information scattering measure. The information affirmation component incorporates two stages: the control message sending stage and information affirmation stage.

All the hubs (Alrajeh, Khan, Lloret, & Loo, 2013) are accepted to have equivalent battery power toward the start. At first, every gateway communicates a neighbor-disclosure message. All hubs in their inclusion zone will answer with an affirmation message. The intrigued bunch head hubs send a group greeting message to every one-jump and two-bounce neighbors to turn into its bunch individuals. One-jump and two-bounce neighbors react back with a bunch of joining transmission. The hub empowers the jump check field. The bunch head gets the group joining tweet. It can affirm that the distance of the joining part is not multiple bounces. Mechanical vibration makes development to electrical energy utilizing piezoelectric, electrostatic, or electromagnetic plans. Photovoltaic cells change over daylight energy into electric form. The hubs work in three states. Latent express, the gateways are effectively taking an interest in activities. It does not collect natural energy. In the semi-dynamic stage, a hub does not effectively be in all the tasks of the organization. It gathers and stores parcels and advances them to the bunch head. The packet is assembled and put away. In the wake of time, it sends the mass bundles. Out of gear state, it does not play out any activity.

The new security steering convention (Kumar, Krishna, & Chatrapati, 2017) is a safe multipath directing convention. There are two kinds of sensor hubs that are desirable over the spot. One is high-design sensor hubs utilized as source hubs host in the organization. The subsequent one is low-arrange sensor devices, called as halfway hubs. These are valuable for multi-hop information transmissions. The organization can fix these hubs to set up the heterogeneous network. It has three significant stages. After sending, the nodes can begin a similar recognizable proof with one another. The hub distinguishing proof organization divides between all the hubs inside the organization. The intermediate node is where the node has a place with the predefined distance factor of the source. It has been characterized and assessed by the convention-based sensor hub correspondence territory. Before sending the encoded form by any confided in the transitional hub from destination to the node, the device confirms their information base for malignant hub addresses. The system cannot send the organization to the gateway. The records are noxious hubs in the transitional hub information base. The format can advance from node to node through the confided in the transitional device with convention. It is having a public key of its sender hub. The public key transmits in the encoded format that assists with scrambling the request during the course disclosure phase. The convention begins the distinguishing proof of securing courses for secure information conveyance. The node can deliver scrambled course demand to the neighbor hubs inside the factor. It can be scrambled with the public key of its neighbor hubs and ship off. The devices can unscramble with their private key and get a store, and update the information. The road hubs replay format and encode. The NSR convention can begin the information conveyance between nodes to the destined node. It is dependent on Elliptic Curve Cryptography.

The sensor hubs (Tang, Liu, Zhang, Xiong, Zeng, & Wang, 2018) separate into five sections: a processor module, a sensor module, a remote correspondence module, a sun-based gatherer, and a battery, the force regulator. The processor, sensor, and correspondence module are equivalent to the modules in a sensor organization. Its sun-based authority, battery module, and force regulator are not the same as those in customary sensor hubs. The sun-based gathers energy-reaping hub model. The capacity changes sun-oriented energy to electrical energy. It happens using photovoltaic impact. The battery is the force supply module. It stores the electrical energy gathered by sun-powered authority and has a restricted limit. The sun-powered can charge it when not energized. When the battery level is full, the sun-powered cannot restore when it gathers more electrical current. The force regulator is the control of the framework. It changes the transmission recurrence of the remote correspondence module. It considers the degree of battery remaining, sun introduction time, and the force of daylight. They receive the improved X-MAC energy utilization model. It has a place with offbeat rivalry MAC conventions. All hubs keep up their obligation cycle in these conventions, and the transmitter and collector are non-concurrent. The getting device may be in rest status when the sending hub sends the information out. Low-energy heading pioneer arrangement will embrace to awaken the accepting instrument. If the sink gets the notification message, it neglects to get the connection information parcel or get the adjusted information bundle. It will locate the malignant hubs assaulted by following the source way of information parcel through the notice with a high likelihood.

The algorithm (Deepa & Latha, 2019) classifies as safe transmission dependent on crossover levels. This calculation is generally valuable in WSN to produce an energy-proficient facilitator hub and a communication information parcel for secure transmission. The energy-effective cycle produces a facilitator Head dependent on which organizer hub has the most elevated force and speed of detecting time. Organizer hub picks the most discussed hub with the closest hub in the organization and is dependent on energy with time. It is a rejection framework based on a credential administration plot. It is named the classified conspire. In this plan, the rekeying message overheads extraordinarily decrease. It subsequently accomplishes energy preservation. It picks the source and objective hubs powerfully, communicating a parcel with information subtleties through coordinating devices. It forestalls the bundle dropping and assault by distinguishing the unsafe devices and sending the data to the facilitator's head. At that point, the chief controller forwards the high-need information to the sink. It calculates the information parcels to be sent to the portable sink. Intermittently facilitator's head checks the line weight. It achieves the edge esteem organizer head promptly and advances its information to sink to maintain a strategic distance from the parcel dropping. The facilitate leader will not check the need for the information parcel. If any hub undetermined in the organization, the arrange hub or facilitate head would dismiss it. It picks another way dependent on trust for the most limited and secure direction. The bundle information conveyed to the objective hub is more effective. It communicates using the chief controller. Energy spent is sent and received. A decent organized head relies upon coordinating device energy content. The arrange hub that has elevated power is the best competitor to be the organized head. The way for sending the data from a group part to the facilitate head is a single-jump, and the Time-division multiple access (TDMA) plan arranges the hub and organizes the head for communicating the data to the versatile sink.

After the organization of sensor hubs, bunch (Maitra, Barman, & Giri, 2019) heads and ways are chosen once dependent on the three boundaries: weight of sensor hubs, signal strength, and way table. The chosen group heads will bite the dust. By running the proposed calculation, it elects the new leader and base station picks them. Bunch heads are with the most worthy weight. The path determines according to the signal strength. The heaviness of sensor hubs relies upon four boundaries of separating from a device to a neighbor hub. Other parameters include the distance between a device and the base station. It also considers the battery intensity, portability, and level of the device. The proposed conspire comprises the arrangement stage, login stage, confirmation stage, key dissemination stage, signature encryption stage, and signature decrypting stage. The key generator creates a transmission message to confirm each hub in the organization intermittently. This transmission message undergoes creation after a specific timeframe to get information about the sensor hubs' actions. It keeps a transmission message counter tally. The estimation is zero for the primary broadcast message, and it will be increased by one for each broadcast message likewise.

The proposed system (Haseeb, Islam, Almogren, & Din, 2019) starts the organization by sending information directions. The directing accomplishes in the beginning stage. The neighbor's data is put away at the nearby table of each hub. The energy-effective grouping utilizes portable bunch heads. It introduces a worldview that holds the close ideal information sending ways toward existing versatile group heads. The subsequent stage presents an information security model to forestall interruption dangers and improves network unwavering quality dependent on blockchain innovation. The information isolates into various squares in blockchain innovation, and all the squares are interconnected employing cryptography standards. They are encoded utilizing a cryptographic hash work, and each square has likewise contained the hash of its past square.

In the proposed energy-productive directing convention (Ferng & Rachmarini, 2012), energy spread hub search fuses and portrays as follows. When the source decides its dispersal hubs, at least one elective method to advance the information to the sink is conceivable. The way determination should be possible with the accompanying contemplations. The source hub considers a spread hub related to the farthest dispersal purpose of the lattice. We essentially consider such a hub the farthest spread hub. The source hub considers the spread hub with the most noteworthy leftover energy as its next jump. Before further explaining when/how to use the two contemplations, it characterizes some assistant documentation. On identifying an upgrade, the source hub produces a parcel. The source hub found close to the sink at that point conveys the bundle straightforwardly to the sink. If the source hub cannot recognize the sink inside its transmission range, the parcel employs the accompanying way. The source hub first checks if it is in the inner area. The halfway device advances the bundle until the message arrives at the sink.

The farthest noteworthy scattering hub search instrument lets the hub locate a shut way from the source to the sink than the others. The energy utilization can be lower, and the organization lifetime delays by adjusting the energy level of sensor hubs in the organization.

The proposition (Mezrag, Bitam, & Mellouk, 2017) depends on two methodologies. The first is cryptography dependent on Elliptic Curves. The subsequent one is the cryptography dependent on the symmetric key for information encryption for MAC activities The contemplated zone is a WSN, in which sensor hubs are homogeneous as far as handling limit, correspondence, energy, and capacity. BS expects to have a limitless asset capacity and is liable for the arrangement before the WSN sending. An assailant should be uninvolved or dynamic during the activity of the organization. BS and the sensor hubs are not versatile. The first is the statement stage. Every device preloads with a key. The setup stage for cycle 0 uses it. It divides among all sent hubs and base stations. The worldwide credential uses for encoding messages to figure a MAC. In the subsequent stage, the bunch head chooses and groups frames where each group head communicates a declaration to neighboring hubs welcoming them to be individuals from its group. The pairwise keys are produced during this stage. The last is the consistent state stage, in which information gathered by the single hubs is sent to CHs and later forwarded by them to the base station. The information traded scrambles, and hence the privacy of transmission is guaranteed. It should notify that the transmitter hub adds a MAC and a nonce in the message shipped off to validation.

For secure steering usefulness in the organization layer, the proposed Self-Channel Observation Trust and REputation System (SCOTRES) (Hatzivasilis, Papaefstathiou, & Manifavas, 2017) coordinates with the Dynamic Source Routing (DSR). Classifying a hub as trusted, real, egotistical, or malevolent is performed after assessing another exchange for this particular hub. The assessment of an exchange's outcome is the fundamental capacity of the convention. It surveys immediate and backhanded information dependent on the SCOTRES's measurements. Backhanded proposals would then be shipped off trusted and authentic one-bounce neighbors, at whatever point the status of an investigated hub changes. It rejects vindictive hubs. The noxious action tends to, and a few assaults counter. The geography metric is determined by each device. It decides the significance of each connected hub for its own directing activity. The primary boundary of the geography metric is the proportion the instrument takes an interest in. It is called Path Support. Hubs with the high investment are significant, as they serve the bundles in numerous ways. The rating segment gets lenient in situations of disappointment. The choice part adjusts the correspondence exertion through elective methods. The edge uncovers the path to restore on the off chance that the assessing hub perceives as vindictive. The second boundary of the measurement is the proportion to objections. It is exceptionally reachable through an inspected instrument called Unique Path Participation. Ousting devices as malevolent will likewise determine these objections inaccessible. The rating segment is lenient toward such a device to forestall misleading indictments in the event of assaults. It rebuffs a malignant device that abuses the measurement after passing the resistance limit. In the steering activity, away from a source to an objective should be chosen. The Path Topological Hugeness determines as the normal Node Topological Significance esteem for every one of way's hubs. The geography metric is a lightweight component with low calculation overhead, as the four boundaries at whatever point steering changes happen. In a regular remote steering convention, the hubs work in unbridled mode to catch the correspondence channel and update their directing information. This instrument advances the data mined by the caught correspondence. It settles on choices about the energy utilization of the assessing hubs. It is dependent on self-perception and not on trading information. Even though the specific energy level isn't determined, the removed data is satisfactory for accomplishing energy and burden adjusting. Neighbor discovery takes place right after the deployment of all sensor nodes. The neighbor discovery is launched at any time by the host during the lifetime of the sensor network. The base station can request to reconstruct this network topology according to the changes of the topology. The base station selects the broadcast key to encrypt the packet neighbor discovery and broadcasts the packet confidential uncover device to the whole network. The fourth step ensures that no secret node discovery packet is announced more than one time for each node, which also applies to other control messages. Thus, the communication overheads for transmitting control packets reduce to a low level. It receives, decrypts, segments, encrypts, and rebroadcasts node discovery. Each node knows its real neighbor and stores them for use in the following phases. The host pauses for a short time to ensure that the base waits for a short time to ensure that the confidential node discovery broadcast floods through the network. Then, the BS broadcasts another packet confidential neighbor collection to collect the neighbor information of each node. The base station sets the current time and the random number to the packet node collection to authenticate each node in the WSN. The broadcast floods through the network. The base station broadcasts another packet confidential neighbor collection to collect the neighbor information of each node. At the same time, the BS sets the current time and the random number to the packet to authenticate each node in the WSN (Kaiwartya, Raw, Abdullah & Cao, 2016). When the sensor node receives the confidential packet, it replies to a validated neighbor collection reply packet to the base by flooding. The session key field, time field, and random number field, the source address sets to itself, and the destination address are set to the base station.

The proposed Ambient Trust Sensor Routing (Zahariadis, Leligou, Voliotis, Maniatis, Trakadas, & Karkazis, 2009) directing convention follows the geological methodology. This methodology is against steering assaults. It considers accounting the productive help of huge sensor organizations. The topographical direction is innately insusceptible against assaults. It identifies with steering message spread, hub ID, and qualities. It is of high significance for steering. The BEACON message reaches out to incorporate the rest of the power field of the source hub. All hubs become mindful of the directions. The excess energy of their neighbors dodges complex computations proposed in the residual power of each neighbor. The energy mindfulness empowers load adjusting is significant for the lengthening of the organization's lifetime and the guard against traffic examination assaults.

When communicating a bundle (Cao, Hu, Chen, Xu, & Zhou, 2006), the sender organizes its neighbors with an assessment capacity and spots this neighbor list in the parcel header. Neighbors on getting the parcel will remember its input for the acknowledged outline and recognize the sender. It meanwhile settles on the autonomous choices of whether to advance the bundle. Criticism from the base station contains the malignant hubs recognized by the base station. The sensor hubs can evade the enemies in directing. To carry the bundle nearer to the objective in an energy-productive manner, the sender will organize its neighbors as per their last time input and put this organized neighbor list in the directing parcel header. The sender communicates the parcel conceding the chosen hub to forward until the cycle of MAC layer dispute.

10.3 ASSUMPTIONS CONSIDERED IN THE PROPOSAL

- The host is the ultimate device that is trustable. It implants procedures, credentials into the small devices before their deployment into the environment.
- Three types of nodes used in the proposal are:
  - The static nodes sense the data, process them, and forward it to the destination through cluster head and different hops.
  - The auxiliary nodes monitor the traffic.
  - The mobile nodes forward the packets
- The nodes opt for different ways to choose their cluster head.

10.4 PROPOSED WORK

The work enhances the previous work (Kumar, Jayasankar, Eswaramoorthy, & Nivedhitha, 2020). The suggestion decreases energy consumption in the nodes and increases reliability. The roles of some of the devices are changed. A new kind of device (auxiliary device) takes care of the traffic and maintains the lookup table. Table 10.1 narrates the work of the proposal.

TABLE 10.1
Working of the Proposal

Step 1: The cluster is formed with the nodes nearby from each other. Similar to the previous work (Kumar, Jayasankar, Eswaramoorthy, & Nivedhitha, 2020), the optimal group chief is elected. The head is randomly selected based on the energy remains (assuming that the network is assault-free) for the first few rounds. If the members come across any assault in the network, they notify the base station.

Step 2: After single instance reporting, the host broadcasts the same to the network. The nodes behave like watchdogs.

The cluster head is selected based on trust value. The cluster member estimates the trust of the group leader. It also calculates the trust of its cluster members. The leader estimates the member's promptness in transmitting packages and the frequency they deliver the packets. The group evaluates using these parameters.

Step 3: Auxiliary nodes are static nodes stationed at equal intervals. They update the lookup table for secure routing. This node transmits the readings to the cluster in regular time intervals.

The rest of the doings are similar to the previous contribution (Kumar, Jayasankar, Eswaramoorthy, & Nivedhitha, 2020). [The system adopts a power-consuming encoding methodology. It generates a secret and shared credential. The cipher uses energy and a confidential key. The head receives the decrypted score and modulus score metric.]

10.5 SIMULATION

The suggestion improves the previous work. It uses assistance nodes. Auxiliary nodes are static nodes stationed at equal intervals. They update the lookup table for secure routing. This node transmits the readings to the cluster in regular time intervals. The work is simulated using NS-2. Table 10.2 provides the details of the same.

10.5.1 Energy Consumption

The work preserves energy in the devices and increases the lifespan of them. It adopts the following measures to conserve power:

- If the continuous monitoring is cut down, the nodes can go to sleep mode (when they are not performing). This methodology conserves energy in them.
- Early assault detection conserves energy in the network.
- The devices of the network choose the cluster head in dual ways. If the network is assault-free, the cluster head election uses the device's remaining energy. If the network is under attack, the trust-based methodology evaluates them, and cluster heads are chosen based on the same. Figure 10.1 represents the comparison amount of power conserved. The suggestion saves 4.038% of energy (Kumar, Jayasankar, Eswaramoorthy, & Nivedhitha, 2020).

TABLE 10.2
Simulation Setup

Parameters Used in the Suggestion	Explanation
Area of surveillance	200 m × 200 m
Number of static nodes in the network	25
Number of clusters formed	5
Number of auxiliary nodes	3
Simulation time	60 seconds

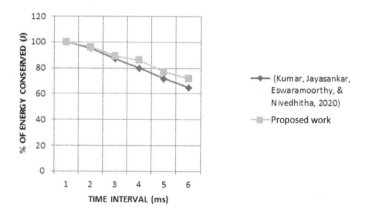

FIGURE 10.1 Comparison of energy conserved.

10.6 CONCLUSION

Sensors are tiny devices used in many applications. Many devices used in the routine of the individual consists of these nodes. They cater to the various needs of the user. The work is an enhancement of the previous contribution. The work minimizes energy consumption and different kinds of attacks by enhancing reliability in the system. The system uses additional devices that share some of the tasks aiding in extending its lifespan. The work increases the energy conservation by 4.038% compared to the previous work. It adapts to the dual mode of cluster head selection. The decision is made based on the assault's existence in the system.

REFERENCES

Ahmed, A., Bakar, K. A., Channa, M. I., & Khan, A. W. (2016). A secure routing protocol with trust and energy awareness. *Mobile Networks and Applications*, 21 (2), 272–285.

Ahmed, A., Bakar, K., Channa, M. I., Khan, A. W., & Haseeb, K. (2015). Energy-aware and secure routing with trust for disaster response. *Peer-to-Peer Networking and Applications*, 10 (1), 216–237.

Ahmed, A., Bakar, K. A., Channa, M. I., Khan, A. W., & Haseeb, K. (2017). Energy-aware and secure routing with trust for disaster response wireless sensor network. *Peer-to-Peer Networking and Applications*, 10 (1), 216–237.

Akyildiz, F.I., Su, W., Sankarasubramaniam, Y., & Cayirci, E. (2002). Wireless sensor networks: a survey. *Computer Networks*, 38, no. 4, 393–422.

Albert, J., Aliu, E., Anderhub, H., Antoranz, P., Armada, A., Asensio, M., et al. (2008). Implementation of the random forest method for the imaging atmospheric Cherenkov telescope MAGIC. *Nuclear Instruments and Methods in Physics Research Section A: Accelerators, Spectrometers, Detectors and Associated Equipment*, 588 (3), 424–432.

Alrajeh, N.A., Khan, S., Lloret, J., & Loo, J. (2013). Secure routing protocol using cross-layer design and energy harvesting in wireless sensor networks. *International Journal of Distributed Sensor Networks*, 9 (1), 1–11.

Ambika, N. (2020a). SYSLOC: hybrid key generation in sensor network. In P.K. Singh, B.K. Bhargava, M. Paprzycki, N.C. Kaushal, & W.C. Hong, (eds), *Handbook of Wireless Sensor Networks: Issues and Challenges in Current Scenario's* (pp. 325–347). Cham: Springer.

Ambika, N. (2020b). Diffie-Hellman algorithm pedestal to authenticate nodes in wireless sensor network. In B.K. Bhargava, M. Paprzycki, N.C. Kaushal, P.K. Singh, & W.C. Hong, *Handbook of Wireless Sensor Networks: Issues and Challenges in Current Scenario's* (Vol. 1132, pp. 348–363). Cham: Springer Nature.

Baoding, Z., & Liu, J. (2010). A kind of design schema of wireless smart water meter reading system based on zigbee technology. *International Conference on E-Product E-Service and E-Entertainment* (pp. 1–4). Henan, China: IEEE.

Cao, Z., Hu, J., Chen, Z., Xu, M., & Zhou, X. (2006). Feedback: towards dynamic behavior and secure routing for wireless sensor networks. *20th International Conference on Advanced Information Networking and Applications-Volume 1 (AINA'06)*. 2, pp. 160–164. Vienna, Austria: IEEE.

Chen, S.L., Lee, H.Y., Chen, C.A., Huang, H.Y., & Luo, C.H. (2009). Wireless body sensor network with adaptive low-power design for biometrics and healthcare applications. *IEEE Systems Journal*, 3 (4), 398–409.

Deepa, C., & Latha, B. (2019). HHSRP: a cluster based hybrid hierarchical secure routing protocol for wireless sensor networks. *Cluster Computing*, 22 (5), 1–17.

Deng, J., Han, R., & Mishra, S. (2006). Decorrelating wireless sensor network traffic to inhibit traffic analysis attacks. *Pervasive and Mobile Computing*, 2 (2), 159–186.

Du, X., Guizani, M., Xiao, Y., & Chen, H.H. (2007). Two tier secure routing protocol for heterogeneous sensor networks. *IEEE transactions on Wireless Communications*, 6 (9), 3395–3401.

Du, X., Xiao, Y., Chen, H.H., & Wu, Q. (2006). Secure cell relay routing protocol for sensor networks. *Wireless Communications and Mobile Computing*, 6 (3), 375–391.

El-Basioni, B.M., El-Kader, S.M., & Abdelmonim, M. (2013). Smart home design using wireless sensor network and biometric technologies. *information Technology*, 2 (3), 413–429.

Fang, W., Zhang, W., Chen, W., Liu, Y., & Tang, C. (2019). TMSRS: trust management-based secure routing scheme in industrial wireless sensor network with fog computing. *Wireless Networks*, 26 (5), 1–14.

Ferng, H.W., & Rachmarini, D. (2012). A secure routing protocol for wireless sensor networks with consideration of energy efficiency. *IEEE Network Operations and Management Symposium* (pp. 105–112). Maui, HI: IEEE.

Haseeb, K., Islam, N., Almogren, A., & Din, I. U. (2019). Intrusion prevention framework for secure routing in WSN-based mobile Internet of Things. *IEEE Access*, 7, 185496–185505.

Hatzivasilis, G., Papaefstathiou, I., & Manifavas, C. (2017). SCOTRES: secure routing for IoT and CPS. *IEEE Internet of Things Journal*, 4 (6), 2129–2141.

Hu, L., & Evans, D. (2004). Localization for mobile sensor networks. *10th Annual International Conference on Mobile Computing and Networking* (pp. 45–57). Philadelphia, PA, USA: ACM.

Kaiwartya O., Raw R.S., Abdullah A.H., & Cao Y. (2016). T-MQM: testbed based multimetric quality measurement of sensor deployment for precision agriculture-a use case. *IEEE Sensors Journal*, 16, (23), 8649–8664.

Khedo, K.K., Perseedoss, R., & Mungur, A. (2010). A wireless sensor network air pollution monitoring system. *International Journal of Wireless & Mobile Networks*, 2 (2), 31–45.

Kumar, K.V., Jayasankar, T., Eswaramoorthy, V., & Nivedhitha, V. (2020). SDARP: Security based Data aware routing protocol for ad hoc sensor networks. *International Journal of Intelligent Networks*, 1, 36–42.

Kumar, K.A., Krishna, A.V., & Chatrapati, K.S. (2017). New secure routing protocol with elliptic curve cryptography for military heterogeneous wireless sensor networks. *Journal of Information and Optimization Sciences*, 38 (2), 341–365.

Lee, H., S., Lee, S., Song, H., & Lee, H.S. (2009). Wireless sensor network design for tactical military applications: Remote large-scale environments. *IEEE Military Communications Conference(MILCOM)* (pp. 1–7). Boston, MA: IEEE.

Lu, H., Li, J., & Kameda, H. (2010). A secure routing protocol for cluster-based wireless sensor networks using ID-based digital signature. *IEEE Global Telecommunications Conference GLOBECOM 2010* (pp. 1–5). Miami: IEEE.

Madria, S., & Yin, J. (2009). SeRWA: A secure routing protocol against wormhole attacks. *Ad Hoc Networks*, 7, 1051–1063.

Maitra, T., Barman, S., & Giri, D. (2019). Cluster-based energy-efficient secure routing in wireless sensor networks. In Chandra P., Giri D., Li F., Kar S., & Jana D. (eds), *Information Technology and Applied Mathematics*. Advances in Intelligent Systems and Computing, (vol. 699, pp. 23–40). Singapore: IEEE. Doi: 10.1007/978-981-10-7590-2_2

Mezrag, F., Bitam, S., & Mellouk, A. (2017). Secure routing in cluster-based wireless sensor networks. *IEEE Global Communications Conference* (pp. 1–6). Singapore: IEEE.

Parno, B., Luk, M., Gaustad, E., & Perrig, A. (2006). Secure sensor network routing: A clean-slate approach. *ACM CoNEXT Conference* (pp. 1–13). Lisboa, Portugal: ACM.

Peijina, H., Tingb, J., & Yandongc, Z. (2011). Monitoring system of soil water content based on zigbee wireless sensor network. *Transactions of the Chinese Society of Agricultural Engineering, 27* (4), 230–234.

Qin, D., Yang, S., Jia, S., Zhang, Y., Ma, J., & Ding, Q. (2017). Research on trust sensing based secure routing mechanism for wireless sensor network. *IEEE Access, 5*, 9599–9609.

Sun, Z., Wei, M., Zhang, Z., & Qu, G. (2019). Secure Routing Protocol based on Multi-objective Ant-colony-optimization for wireless sensor networks. *Applied Soft Computing, 77*, 366–375.

Tang, J., Liu, A., Zhang, J., Xiong, N.N., Zeng, Z., & Wang, T. (2018). A trust-based secure routing scheme using the traceback approach for energy-harvesting wireless sensor networks. *Sensors, 18* (3), 1–44.

Xiangning, F. (2007). Improvement on LEACH protocol of wireless sensor network. *International Conference on Sensor Technologies and Applications* (pp. 260–264). Valencia, Spain: IEEE.

Yang, T., Xiangyang, X., Peng, L., Tonghui, L., & Leina, P. (2018). A secure routing of wireless sensor networks based on trust evaluation model. *8th International Congress of Information and Communication Technology (ICICT-2018). 131,* (pp. 1156–1163). ScienceDirect.

Yin, J., & Madria, S. K. (2006). A hierarchical secure routing protocol against black hole attacks in sensor networks. *IEEE International Conference on Sensor Networks, Ubiquitous, and Trustworthy Computing (SUTC'06). 1,* (pp. 8–16). Taichung, Taiwan: IEEE.

Zahariadis, T., Leligou, H. C., Voliotis, S., Maniatis, S., Trakadas, P., & Karkazis, P. (2009). Energy-aware secure routing for large wireless sensor networks. *WSEAS Transactions on Communications, 8* (9), 981–991.

11 Deploying Trust-Based E-Healthcare System Using Blockchain-IoT in Wireless Networks

Amanjot Kaur
IKGPTU

Parminder Singh
CGC Landran

CONTENTS

11.1 Introduction 189
    11.1.1 A Brief State of the Art in Terms of Study Hypotheses 190
11.2 Related Work 193
    11.2.1 International Status 194
    11.2.2 National Status 195
11.3 Blockchain Technology in E-Healthcare System 195
    11.3.1 Different Aspects of Blockchain 196
    11.3.2 Features of E-Healthcare System Using Blockchain 196
    11.3.3 Working Principle of Blockchain 196
    11.3.4 Importance of the Proposed Paper in the Context of the Current Status 197
11.4 Technical Details of IoT-Blockchain for E-Healthcare System 197
    11.4.1 Model Implementation 199
11.5 Conclusion 199
References 200

11.1 INTRODUCTION

The e-healthcare system is likely to solve the most significant problems associated with the healthcare system, including lower costs and greater access to care. The Health Information Technology for Economic and Clinical Health (HITECH) Act was signed into law on February 17, 2009 [1], leading to significant investments in health IT. The problem of data breaches has always been theirs in the healthcare system. According to the survey, 50.2 million data leaks are expected in 2021 [2]. The exchange of health information brings information about patients, clinical trials, and the care team. To be accurate, the patient is placed in the center of the sphere and around all sources to exchange health information. It accurately collects data on medication, clinical outcomes and other care settings. A recent survey predicts that newer models and Electronic Health Records (EHRs) systems will need to be more effective for information exchange and interoperability systems. The exchange of information plays a vital role in improving the health of the population, including clinical trials, monitoring the safety of medical products, detecting, and responding to medical practices. To share information, there must be an interaction between EHRs and other clinical software. It has been observed that the condition reduction centers acquired by the hospital represent more than 96% of the center for Medicare and Medicaid (CMS) which adopt EHR, but these EHRs are not used for direct measurement [3]. Hospitals are needed and health systems are moving to real-time EHRs data. EHR vendors and other healthcare IT vendors use state-of-the-art technology to collect and analyze patient safety data and provide feedback to hospitals and clinical teams without retribution. Health Information Exchange (HIE), health information is estimated by electronic devices such as clinical summaries and laboratory results.

Analysis of HIE under the following categories: disease monitoring, patient experience (satisfaction), and use of healthcare (including reductions and unnecessary lab tests). HIE results are based on positive and negative feedback. The goal of HIE is to reduce electoral bias and deliver better results in the right way [4]. Research articles found that patients had strong support for self-administered medications and mixed-quality evidence to support patients with chronic disease self-management, misdiagnosis, and anticoagulation medications with accuracy and accurate medical records. Promises of patient involvement in safety, such as anticoagulation management and patient portal access, are not widely implemented. However, it must also be acknowledged that patients' preferences, self-efficacy, and power dynamics must be based on priorities in safety initiatives [5]. Another patient engagement study requires the involvement of patients, families, and carers to improve healthcare safety. Determining patient involvement is a separate study and research needs to be done. The following criteria appear to be administrative errors; some technologies have corrected the negligence of patient monitoring and schedule for appointments. The article found that no patient safety studies were conducted based on the involvement of other members of the care team, including family care providers, home healthcare providers, pharmacists, and nurses. It was also found that there is no patient safety based on the partnership between health medical and social drivers [5]. The World Health Organization (WHO) sheds light on guidelines for patient safety engagements. Therefore, in our chapter, we review the guidelines and consider enabling the perfect system for patient engagement.

11.1.1 A Brief State of the Art in Terms of Study Hypotheses

Conventional security methods are not practical for Internet of Things (IoT) because IoT topologies are largely centralized and the IoT devices have limited resources. Furthermore, IoT is the technology to gather data through sensors in a centralized manner. The function of IoT devices is to collect patient data and evaluate those data on a periodic basis. The patient who does not want to go to hospitals can be monitored at home. Real-time remote patient monitoring (RPM) has been enabled through IoT system.

blockchain has been used to provide security in a similar environment used for cryptocurrencies. A blockchain is immutable, meaning it is unable to be changed. Many industries are looking at blockchain technology to securely track transactions and quickly provide accurate information regarding those transactions. Blockchain operates as a peer-to-peer architecture, and the peer is not owned by Internet service providers (ISPs). Peer communicates messages without being passed through the dedicated server. Most instant applications track the IP address of a machine through a server, but peer-to-peer (P2P) sends a message directly to another user without being tracked of IP address. Each peer workload generates the required file, adding service capabilities to each peer system. The blockchain model can overcome the three main challenges of P2P models.

- The model is not dependent on ISP; it requires a minimum amount (or transaction) assigned to another block.
- The hash algorithm is used in blockchain models to enhance the security features of the blockchain model.
- The blockchain peer requires minimal space and bandwidth to process the block. The processed block uses upstream traffic if downstream traffic is not available.

Blockchain work like peer-to-peer architecture and peer did not own by the ISPs. The peer communicates without passing the message through a dedicated server. Many instant applications access the machine's IP address through servers, but in the P2P message sent directly to another user. Each peer generates the workload requested file, and each peer also adds the service capacity to the system. The blockchain model can overcome the three main challenges of P2P models.

- The model is not dependent on ISP; it also requires a minimum amount of data (or transaction) delivered to another block.
- The hash algorithm used in the blockchain model enhances the security features of the blockchain model.
- Blockchain peer requires a minimum amount of space and bandwidth to process the block. The processed block utilizes upstream traffic if downstream traffic is unavailable.

Although blockchain is primarily known as the technology behind Bitcoin, it is gaining a lot of attention from those interested in finding better ways to secure any type of transaction, including information exchanged by IoT devices.

Different types of blockchain support managed data, including public permissionless, consortium (public permission), and private [6]. Blockchain of the public permission-less type is accessible by everyone and only part of the data is encrypted. A Node may join with minor or without permission. The consortium works on a set of nodes to participate in the allocation process. It is also adopted by the industry and is open to the public. A consortium blockchain can be formed as a combination of public and private blockchain. This type of blockchain handles transactions for a selected number of users. It also features cryptography hash functions that are found in public blockchain. The overall functioning of this blockchain is similar to that of a semi-decentralized model, which means that the number of selected users controls the total number of participants. In a central way, this type of blockchain is used in cross-discipline solutions. It also improves impact and transparency in the workplace. This is a partly central trust approach. The last type is a private blockchain that can be applied to a network and joined by miners or nodes. Private blockchain knows who can join the network and assign different roles to the users. It is easy to control the users, and therefore, throughput is high in this case. They perform minor or node transactions and operate under smart contracts. It is used for private purposes but still believes in the third party.

The following diagram represents the blockchain model which is used in blockchain types. The block is mined and added to the longest chain. When the blocks are propagated, a new chain is forked, if block is not part of the longest chain, which is always preferred in the network, that block is called an orphan block. The miner neglects these blocks, every new blocks built, and part of longest chain. Block represents the series of transactions. It may grow up to 8 MB or sometimes higher. Orphan block is not a part of the longest chain and these blocks further not being used in transaction help in processing a large number of transaction in one go. The structure of the blockchain consists of a block header and list of transactions. The block header started from block number and at the bottom of the block contains a list of transactions illustrated in Figure 11.1. The block header is connected to series of blocks and every block inherits the previous block. We use the previous block to create the new hash.

So far, the nominal focus of research is on domains like IoT, blockchain, and healthcare systems. Blockchain is a decentralized computation made up of multiple transactions and each transaction retrieves the previous block information. The healthcare system is another domain that affects human life. The integration of blockchain and IoT technology provides many opportunities for the healthcare system [7]. The healthcare blockchain must address many technological, behavioral, and economic challenges before being adopted by a nationwide organization.

The working model of IoT-blockchain is shown in Figure 11.2. The major contribution of this study can be summarized as follows:

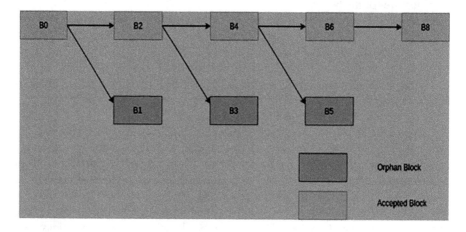

FIGURE 11.1 Blocks created in blockchain model (longest chain).

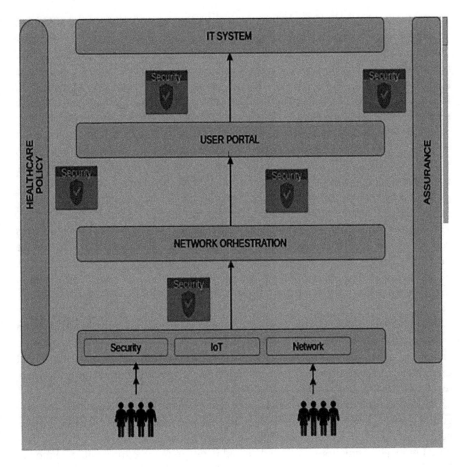

FIGURE 11.2 Working model of IoT-blockchain model.

- Providing a quality assurance model for developing digital e-healthcare systems.
- Designing a medical application to engage patients and doctors.
- Implementing security architectures in the digital system to protect the confidentiality of medical devices, information dissemination, and patients' databases.

11.2 RELATED WORK

Blockchain technology is a great tool in key sectors such as finance and banking. Blockchain focuses on security and certified decentralized fashion without third-party concerns. Arrange the data in blocks that cannot be edited and the transaction history is added to the chain structure. Block chain has the potential to transform healthcare systems [1], placing the patients at the center of the healthcare ecosystem and increasing security. Privacy and interoperability are the two coins that make up an efficient healthcare system. Blockchain maintains a country's large-scale health registry that can facilitate accountability and examine key points of improvement for the supply chain. Identifying new accidents, viruses complement HINI and reduce human damage. Thirty percent of the studies provided a mechanism for managing healthcare records, while none provided actual large-scale validation in practice [2]. IoT networks and blockchain technology aim to eliminate privacy and security threats to data integrity. Smart contracts used in models that have smart contracts between peers. Smart contracts can handle authentication, authorization, and access control and data management. Node violations can cause network abnormalities but the cryptography hash function can store our data in blocks. Central security models are not possible for large-scale networks and, therefore, distributed networks provide a new provision for securely transmitting data using cryptography functions. Authentication protocols are lightweight, but they do not provide protection. The overall framework template presented by the authors [9] works in three layers which can lead to user and database security. Blockchain is required to authenticate integrity data with sensitive data with timestamps. Various conceptual architectures have discussed edge-based IoT medical record systems [10], which is insufficient in the security point of view. Medical procedures are needed for disabled and elderly patients [4].

IoT-enabled healthcare devices investigate ubiquitous computing [12] which involves different systems like human to human (H2H), human to machine (H2M), and machine to machine (M2M). In [13], the authors implemented a prototype model that collected the data in the form of pulse rate and body temperature. The user receives a notification through the Gmail application.

All the measurements are stored in the IoT platform. On the downside, the suggested model is bulky and lacks security. Authentication model in [14] has derived the patient data collection and transfer through GSM module to Azure IoT hub. The proposed model can monitor real-time data and is integrated into the healthcare model. It is also possible to remotely monitor the health through various devices like sensors and GSM module. Data is acquired with the help of a server. The encryption methods to position the fuzzy interface controller and the developed architecture reduce the overhead of the accessibility time. The proposed model is based on IoT home care information systems [15], designed for internal and external environments. The suggested method is to keep care of elderly people who have limited mobility and are living alone. Suggested Inertial measurement unit (IMU) system equipped with WiFi, GPS, and heart rate modes and validation procedures performed on crucial activity recognition and localization requirements.

11.2.1 International Status

It is difficult to ascertain the effects of cyber-attacks, which can have an economic impact on any country. According to the reports, the estimated loss from these attacks is $500 a year. Similar attacks have occurred in the medical field as Cybercriminals obtain a list of Personally Identifiable Information (PII). In PII, there is information such as name, social security number, credit card, address information, and more. Cybercriminals get information and sell it in the dark web. Including thousands of nodes badly deploy in a healthcare network can increase the likelihood of weakness as it connects thousands of unlocked doors. The threat actor may exploit the health system using these unlocked doors. Weak authentication, embedded server processes, and possible vulnerability to embed user processes are also caused by user-made errors. The United States [3] and the European Union are very concerned about security issues as the health departments have also seen data breaches. Ambient Assisted Living Joint Programme (AALJP) is a precise ample European project funded [16], and they continuously demand research proposals related to health systems. However, the World Health Organization (WHO) gives timely guidelines to the United Nations (UN) on health issues. The organization also wants to make a standardization system to solve health services problems to provide good services to the patients.

11.2.2 National Status

In India [12], the ICT-based health system adopted by the government was promoted over the decades. Reference information is also available on the website of Health department. Several new health monitoring systems have been developed to assure accurate data monitoring of patients, such as a raspberry pie system equipped with health sensors [17]. In another report, the physical ID information associated with the Radio Frequency Identification (RFID) tag IoT is obtained. This device also allows doctors to access patient information [18]. HealthFog technology has been suggested by the authors that collect the information through with the help of IoT devices. It improves the quality of service issues, which immediately falls due to slow Internet connectivity. The suggested improves the overall performance.

According to the literature discussed above, some technologies are still being started, or are happening. Some of the technologies are being explored by developed countries and have not been implemented in our view. The current state of this domain needs to be enhanced to provide better services with better modifications to existing infrastructure. In the meantime, weakness is always in the medical equipment as everything is connected to the Internet, leading to many risks.

11.3 BLOCKCHAIN TECHNOLOGY IN E-HEALTHCARE SYSTEM

Blockchain overcomes the limits of traditional accessibility, including lack of security and centralization [7,8,11]. Blockchain multiple transactions are created by Bitcoin, retrieving information from the previous block that cannot be exchanged or deleted. For example, a blockchain is a distributed database, meaning it is stored in different computers in a P2P fashion. We say that blockchain is evidence of tampering; every transaction in the block is legitimate and secure. Every transaction on the blockchain is anonymous to the printer and counted by the address of the public key concept and wallet applications. The wallet hears the transaction address in a user account. A valid user uses each address in Bitcoin. So the transaction speed is also encrypted and the enabled target node can decrypt the transaction and accept it. However, the actual transaction amount is open to everyone for verification.

Healthcare is one paradigm and has a lot of scope in clubbing with blockchain technology [19]. The concept of blockchain works on consensus. This process agrees to distribute nodes (like miner) to reach a final state of data. Consensus provide the insecurity of reliability and fault tolerance in the blockchain networks. Each transaction will take place in a distributed way and distributed protocols work together in the network. The US healthcare provides digital services to the customers at their own end to realize the full potential in e-healthcare and to shape in blockchain future [20].

11.3.1 Different Aspects of Blockchain

Nonce: It is an arbitrary number that will be used only once. This number ensures that the previous communication could not be part of a malicious attack.

Paper Wallet: It is a document that contains the necessary information to access a certain amount of Bitcoin.

Protocol for Commitment: It ensures that every valid transaction from the client is committed and included in the blockchain within a finite time.

Consensus: It ensures that the local copy is consistent and updated.

Security: Data that move from one block to another is tamper-proof. If the node misbehaves in the transaction, it simply identifies the block of the previous hash value and cannot add in the blockchain.

Privacy and Authenticity: The data belong to various clients so privacy and authentication kept in the blockchain.

11.3.2 Features of E-Healthcare System Using Blockchain

Features of e-healthcare are as follows:

a. Decentralization: This is the process of spreading data across blocks and away from the central authority. This process requires an e-healthcare system and does not allow authentication data to be leaked. This means that the owner has complete control over his own transactions. Decentralized applications that operate in P2P networks typically use blockchain and smart contracts.

b. Disintermediation: Blockchain is designed to eliminate intermediate or middlemen in a transaction. Therefore, the intermediate doesn't access the information in the blockchain network. The dealt transaction is tamper-proof, which means the owner not worry about the data loss. Bitcoin, part of the blockchain, enables the exchange of value between two parties on the Internet.

c. Permissioned ledger: Transactions can only be authenticated and processed by a participant who is already recognized and authorized by the regeneration authority. Permitted ledgers are usually associated with private blockchain.

11.3.3 Working Principle of Blockchain

The consensus process was discovered by Satoshi Nakamoto [21]. It is decentralization that provides a distributed layout in which services are provided to the customer. Four different processes occur independently across the network.

- Verification of transactions through nodes.
- Adding traffic to a new block by mining nodes, displaying proof of work.
- Verification of new blocks by each node and assembled in a chain.
- Selecting each node of the chain with maximum calculation by proof of work.

Transactions are pseudo-anonymous, providing public key concept and addresses generated by crypto keys. A valid user verifies every address in Bitcoin. For address verification, the wallet plays an important role in storing digital currency such as Bitcoin and Litcoin. The target node can decrypt the transaction and accept it. However, the transaction amount is publicly open for verification. This makes verification easier. The miner chooses the best chain in the blockchain which will be applicable for the e-healthcare model; from our point of view, the longest chain is the accepted chain. If the length of the chain is the same, then the miner will prefer the selection of the best chain.

11.3.4 Importance of the Proposed Paper in the Context of the Current Status

IoT-blockchain provides many functions, including security and provides a platform to integrate many medical devices at one time. In the meantime, the proposed method is a one-way step for data collection and monitoring of patients. The proposed interface creates a direct connection between the patient and the medical representative. Despite these features, a feature will also be developed to help us track the status of medical devices. Our models assure customers that a system that will be developed to better the health system can help prevent equipment failure.

11.4 TECHNICAL DETAILS OF IoT-BLOCKCHAIN FOR E-HEALTHCARE SYSTEM

In the e-healthcare system, the major problems facing the healthcare systems are inaccuracy of data, scalability, and security [22,23]. Health appliances are not compatible with IoT devices, and there is a demand for more development in this area [24,25]. Many enterprise-level servers are on the market, but they cannot integrate two technologies. Therefore, the model's existing problem is accurate monitoring, reporting, and troubleshooting capabilities.

Figure 11.3 presents a technical model of the integrated system, which includes multiple technologies. For this perspective, IoT gateways play an important role in the given architectural framework. The IoT gateway collects data from IoT-enabled devices and forwards the information to the healthcare system. It provides the patient interface from different applications that are supported by devices like IoT sensor, wearable devices, smart phones, etc. Medical professionals diagnose the problem and give a prescription to the patient and that information is stored in the blocks. The blockchain manages each block, and the data is stored in the encrypted form. Medical professionals cannot obtain patient information. The following equipment is required to complete the research project.

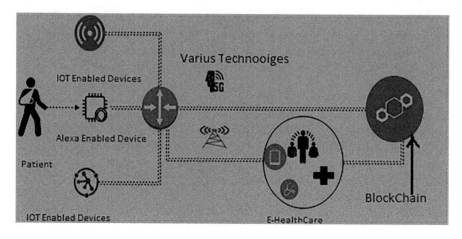

FIGURE 11.3 Technical framework of IoT-blockchain model.

1. Router
2. Switch
3. IoT Gateway
4. IoT Devices (sensor, wearable band, smart watch, etc.)
5. Software (support IoT devices, preferably built own application)
6. Blockchain OS

The proposed model is a really sophisticated policy-based access control system, as shown in Figure 11.3. It can also be used for medium to large-scale platforms. The simple network management (SNMP) protocol is integrated into IoT devices and can control blockchain. It can also elaborates the status of each report by email to the patient. Flexible and detailed service administration in IPv4 and IPv6 networks is also available in this model and has the required full audit and reporting capabilities.

The following properties of IoT-blockchain can help to tackle major problems in the healthcare system:

- Electronic Medical Record: The blockchain platform supports the entire life cycle model of patients' medical records. It guarantees safe patient records, including billing documentation, medical reports, and more. It provides robust security and impeccable audit-ability.
- Clinical Test: Blockchain can provide accountability and transparency in the clinical trial's reporting process by detailing all trials. These clinical trials may also indicate which medical approaches work best for certain illnesses or groups of people. Clinical trials produce high-quality data for healthcare decisions.
- Internet of Things (IoT): IoT provides a variety of functionalities for medical devices. In addition to monitoring and presenting patient statistics, IoT sensors can be used to track the location of these devices.

Deploying Trust-Based E-Healthcare System

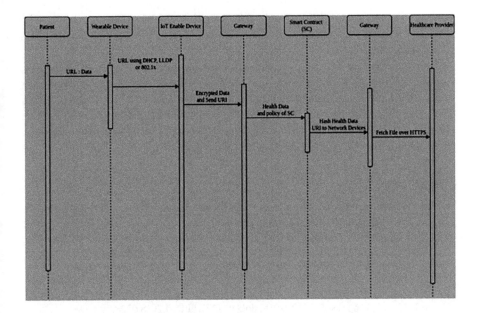

FIGURE 11.4 Proposed model of IoT-blockchain model.

11.4.1 Model Implementation

Smart contract is an autonomous agent and it executes by itself. It is P2P agreement that live on blockchain forever. The smart contract removes the trustworthiness of the third party and remains active in the P2P fashion. It provides a strong relationship in patients and healthcare providers. The other functionality of this serverless contract is to establish authentication between the two peers. In the implementation model, shown in Figure 11.4, the patient is connected through IoT-enabled device through Uniform Resource Locator (URL). IoT allows the device to provide IP addresses using dynamic host configuration protocol (DHCP). These IoT-enabled devices are connected to the gateways. The health data is encrypted and if rules are met, then the smart contract is automatically executed. Smart contracts send hash Uniform Resource Locator (URI) through networking devices. On the other end, healthcare providers fetch the file over Hypertext Transfer Protocol Secure (HTTPS).

11.5 CONCLUSION

The healthcare industry focuses on important proposals that provide better healthcare and personalized and cost-effective care. However, the healthcare industry faces many major concerns in e-healthcare industry due to the lack of medical specialties, rising healthcare costs, and safety breaches. Blockchain-IoT healthcare provides better security, reduces costs, and improves productivity. This system provides care everywhere, in rural and urban areas. In this technique, the patient can be seen almost and being aware of the situation is the key to healthcare. Blockchain enables enterprises to build a trust-based business network for digital transformation. This chapter discussed the blockchain-IoT technologies that can improve healthcare systems. The proposed framework integrates IP-based networks and provides network access across large networks.

The framework provides a way to interact with any input such as email, messaging, data transfer through wearable devices. In the next use-based scenario, a smart contract was created between the patient and the healthcare provider. Use case uses consensus to include data in blockchain; from some agreements, data is formed in encryption and stored in blockchain. In future studies, the Blockchain-IoT-based system could be implemented in a real-time perspective to make it accessible to large networks.

REFERENCES

1. E. H. Benefits, "Health policy brief," *Health Aff.*, no. Cdc, 2015, [Online]. Available: http://www.healthaffairs.org/healthpolicybriefs/.
2. M. H. Kassab, J. DeFranco, T. Malas, P. Laplante, G. destefanis, and V.V.G. Neto, "Exploring research in blockchain for healthcare and a roadmap for the future," *IEEE Trans. Emerg. Top. Comput.*, vol. 6750, no. c, pp. 1–16, 2019, doi: 10.1109/TETC.2019.2936881.
3. D. Classen, M. Li, S. Miller, and D. Ladner, "An electronic health record–based real-time analytics program for patient safety surveillance and improvement," *Health Aff.*, vol. 37, no. 11, pp. 1805–1812, 2018, doi: 10.1377/hlthaff.2018.0728.
4. S. Rahurkar, J. R. Vest, and N. Menachemi, "Despite the spread of health information exchange, there is little evidence of its impact on cost, use, and quality of care," *Health Aff.*, vol. 34, no. 3, pp. 477–483, 2015, doi: 10.1377/hlthaff.2014.0729.
5. A. E. Sharma, N. A. Rivadeneira, J. Barr-Walker, R. J. Stern, A. K. Johnson, and U. Sarkar, "Patient engagement in health care safety: An overview of mixed-quality evidence," *Health Aff.*, vol. 37, no. 11, pp. 1813–1820, 2018, doi: 10.1377/hlthaff.2018.0716.
6. M. Hölbl, M. Kompara, A. Kamišalić, and L. N. Zlatolas, "A systematic review of the use of blockchain in healthcare," *Symmetry (Basel).*, vol. 10, no. 10, 2018, doi: 10.3390/sym10100470.
7. P. P. Ray, D. Dash, K. Salah, and N. Kumar, "Blockchain for IoT-based healthcare: background, consensus, platforms, and use cases," *IEEE Syst. J.*, pp. 1–10, 2020, doi: 10.1109/jsyst.2020.2963840.
8. T. McGhin, K. K. R. Choo, C. Z. Liu, and D. He, "Blockchain in healthcare applications: Research challenges and opportunities," *J. Netw. Comput. Appl.*, vol. 135, no. February, pp. 62–75, 2019, doi: 10.1016/j.jnca.2019.02.027.
9. K. P. Satamraju and B. Malarkodi, "Proof of concept of scalable integration of internet of things and blockchain in healthcare," *Sensors (Switzerland)*, vol. 20, no. 5, 2020, doi: 10.3390/s20051389.
10. A. F. Subahi, "Edge-based IoT medical record system: requirements, recommendations and conceptual design," *IEEE Access*, vol. 7, pp. 94150–94159, 2019, doi: 10.1109/ACCESS.2019.2927958.
11. J. Abdollahi, B. Nouri, M. Mehdi, and E. Parvar, "Improving diabetes diagnosis in smart health using genetic-based ensemble learning algorithm approach to IoT infrastructure," *Futur. Gener. Distrib. Syst. J.*, vol. 1, no. 2, pp. 26–33, 2019.
12. A. A. Albesher, "IoT in health-care: recent advances in the development of smart cyber-physical ubiquitous environments," *IJCSNS Int. J. Comput. Sci. Netw. Secur.*, vol. 19, no. 2, pp. 181–186, 2019, [Online]. Available: https://www.researchgate.net/publication/331642487.
13. N. M. Ying, "Chapter 2 measurement of body temperature and heart rate for the development of healthcare system using IoT platform," *Bioeng. Princ. Technol. Appl.*, vol. 2, pp. 15–25, 2019.

14. H. A. El Zouka and M. M. Hosni, "Secure IoT communications for smart healthcare monitoring system," *Internet of Things*, no. xxxx, p. 100036, 2019, doi: 10.1016/j.iot.2019.01.003.
15. D. Dziak, B. Jachimczyk, and W. J. Kulesza, "IoT-based information system for healthcare application: Design methodology approach," *Appl. Sci.*, vol. 7, no. 6, 2017, doi: 10.3390/app7060596.
16. M. T. Mardini, Y. Iraqi, and N. Agoulmine, "A survey of healthcare monitoring systems for chronically Ill patients and elderly," *J. Med. Syst.*, vol. 43, no. 3, 2019, doi: 10.1007/s10916-019-1165-0.
17. D. M. J. Priyadharsan, K. K. Sanjay, S. Kathiresan, K. K. Karthik, and K. S. Prasath, "Patient health monitoring using IoT with machine learning," *Irjet*, vol. 6, no. 3, pp. 7514–7520, 2019, [Online]. Available: www.irjet.net.
18. V. S. Naresh, S. Reddi, N. V. E. S. Murthy, and Z. Guessoum, "Secure lightweight IoT integrated RFID mobile healthcare system," *Wirel. Commun. Mob. Comput.*, vol. 2020, pp. 1–13, 2020, doi: 10.1155/2020/1468281.
19. C. Agbo, Q. Mahmoud, and J. Eklund, "Blockchain technology in healthcare: a systematic review," *Healthcare*, vol. 7, no. 2, p. 56, 2019, doi: 10.3390/healthcare7020056.
20. A. Agrawal, "A study on integration of wireless sensor network and cloud computing: requirements, challenges and solutions," *ACM*, 2015, pp. 35–40.
21. S. Nakamoto, "Bitcoin: a peer-to-peer electronic cash system," 2008. doi: 10.1162/ARTL_a_00247.
22. P. Sing and R. S. Raw "Development of novel framework for patient health monitoring system using VANET: an Indian perspective", *Int. J. Infor. Technol.*, Springer, 13, 2021, pp. 383–390.
23. P. Gaba and R. S. Raw, "The amalgamation of blockchain with smart and connected vehicles: requirements, attacks, and possible solutions", *IEEE International Conference on Advances in Computing, Communication Control and Networking (ICAC3N-20)*, Greater Noida, 2020, India, Doi: 10.1109/ICACCCN51052.2020.9362906.
24. S. Kumar and R. S. Raw, "Health monitoring planning for on-board ships through flying ad hoc network," *Adv. Comput. Intell. Eng.*, Springer, 2020, pp.391–402.
25. P. Singh and R. S. Raw, "State of the art in simulation software for health monitoring through vehicular ad-hoc network," *6th International Conference on Computing for Sustainable Global Development (INDIACom-2019)*, New Delhi, 2019, pp. 832–837.

12 Low Cost Robust Service Overloading Fusion Model for Cloud Environments

Sitendra Tamrakar
Nalla Malla Reddy Engineering College

Ramesh Vishwakarma and Sanjeev Kumar Gupta
Rabindra Nath Tagore University Bhopal

CONTENTS

12.1 Introduction 203
12.2 Cloud Computing Environment Security Issues 204
12.3 Objectives and Significance..205
    12.3.1 Objective 205
12.4 Fusion Model 206
    12.4.1 Cloud Overloading..207
    12.4.2 Service Overloading 208
12.5 The Proposed Model...208
12.6 Overloading Authentication System 209
12.7 Implementation and Results 210
    12.7.1 The Simulation Environment.. 211
    12.7.2 Simulation Results and Analysis 211
12.8 Conclusion 215
References... 216

12.1 INTRODUCTION

Cloud Computing is the process of assembling and administrating computing services on stipulations. The services provided by the cloud are an amalgamation of applications, knowledge centers on diverse networks, and a pay-as-you-click feature provided by the cloud to their consumer. The feature of cloud computing has frequently been called as "Cloud" or "cloud to Cloud" [1]. The integral process of assembly and distribution services adds diligent worth to industry, maneuvered and modernism, and also there are barriers to cloud implementation in areas related to security space. The security considerations cause the downside of trust [2–5] with the cloud.

Using this theoretical information like execution complexity, network attacks, Wireless (cordless)-Local-Area-Network (WLAN) security, advantages and disadvantages aren't sufficient to choose a distinctive authentication method. Thus, experimental network analysis is required to choose cloud security [6,7] for authentication, which gives a preferable performance in terms of secure authentication time and total processing time (TPT). This prompts us to evaluate the network performance of widely used cloud security techniques [8–10] for both wired and wireless networks. We calculated two parameters, namely, communication cost and authentication time.

It is compulsory to compare the scalability of both wireless and wired cloud networks in the real world. This scalability will give us the information if the same protocol can be used for both wired networks and wireless cloud networks, or different protocols need to be used for wireless and wired networks.

With the rising utilization of cloud services, cloud computing has augmented safety services as the key issue in a very cloud computing atmosphere. Customers don't need to lose their non-public data as a result of malevolent informants and outsiders and from any terminal calamity of cloud services. Additionally, the loss of service handiness has caused several issues for several customers recently. The objective of current work intended a replacement model known as Service Overloading Fusion Model Sharp that uses Shamir's secret sharing rule with service overloading technique rather than single cloud or multi-cloud suppliers.

12.2 CLOUD COMPUTING ENVIRONMENT SECURITY ISSUES

In the previous couple of years, cloud technology has grown up from being promising information and knowledge security drawback and business notion to one of the growing IT business areas. This analysis section highlights the protection issues and needs that stem from the characteristics properties of cloud surroundings and propose a technical description of the foremost relevant attack eventualities.

Reliable service and secure information availability: Secure network dependability may be a key cornerstone for cloud computing environments and cloud services. As a result of a cloud being retrieved over public native systems, the cloud suppliers should illustrate probable forfeiture of web network spinal column property. The common drawback ought to be the most thought for a cloud service user WHO entrusts infrastructure to the cloud. Availableness is additionally the main drawback for personal cloud infrastructures.

Security: wherever is your information safer? On your native storage (personal storage), or on secure servers within the cloud surroundings? Some argue that user information is additionally secure once managed inside, whereas alternatives infer that cloud suppliers (SP) have a robust stimulant for higher security. But within the cloud environment, your information and knowledge are contacted these individual computers notwithstanding wherever your base repository of knowledge is ultimately stored- assailant will attack just about any native server or cloud server and some statistics show that third (OT) of breaches result from the purloined portable computer and other devices and from staff accidentally showing data on the net with nearly Sixteen Personality Factor Questionnaire thanks to corporate executive theft.

Privacy: Cloud computing build based on the virtualization technology, cloud users, personal information could also be extended in several data center (DC) instead of keep within the same location, even across international borders at this time information privacy protection [8,9] can face the dispute of various legal systems. On the opposite side, cloud customers could leak hidden information or data once they are accessing (use) cloud services. Cloud attackers will analyze task based on computing task submitted by the cloud users.

Cost-effectiveness: The significant aspects employed through CSP to market their elucidations of price but effort whole software/ hardware design.

The work done antecedently [1,11–13] focuses chiefly on theoretical aspects, and less experimental work has been done concerning information and cloud services. Exploitation of this theoretical data like implementation quality, network attacks, Wireless native space Network (WLAN) security, advantages, and drawbacks isn't enough to decide on a selected authentication technique. Hence, AN experimental analysis is needed to decide on cloud security for authentication, which supplies higher performance in terms of authentication time and total interval. This intended America to gauge the performance of wide used cloud security ways for each wired and wireless network. We have a tendency to calculate two parameters particularly, total interval and authentication time.

In the time period, it's necessary to match the measurability of each wired and wireless cloud network. This data can provide America the information if the identical protocol is used for each wired network and wireless cloud network, or totally different protocols have to be used for wired and wireless networks.

To get the opinion of scholars regarding the cloud information security provided by Rabindra Nath Tagore University (RNTU) field, a user survey was chosen. The motive behind the survey was to grasp if the scholars can look ahead to few additional seconds to induce higher security with regard to authentication. It additionally attention-grabbing to grasp the opinion of the scholars concerning the protection they need been exploited at the BTH field as they're the end-users.

12.3 OBJECTIVES AND SIGNIFICANCE

12.3.1 Objective

In this analysis work, we've tried to develop a model to enhance the Security of Hybrid Cloud within cloud computing. If the model enforced in a high level and established in contradiction of factors to make sure the services overloading by benefiting from securing cloud.

The following are the aims of the planned learning in specific terms:

- To mix recent security techniques.
- To outline parameters for Fusion Model.
- To opt for measurability price.
- To enhance security techniques on giant information.

12.4 FUSION MODEL

Cloud computing is the utmost dynamic sector that has been developing at a tremendous ratio in the past few years. The significant module of this evolution and expansion of the cloud is the amenities delivered by the artifact that provide distinct storage and computational features at much economical cost, preferably favored by the software companies and IT industries. Despite all the assurance and potential proven by the cloud service supplier or provider, the major concern remains regarding the security of information of the client who has been offered the cloud services. These issues arise due to the outsourcing of data via third-party cloud system platforms. These security vulnerabilities make cloud services less versatile. Our work proposes a secure service overloading Fusion Model that enables knowledge to be held on firmly within the cloud whereas at identical time permitting operations to be performed thereon with no compromise of the sensitive elements of the information.

Cloud may be a service field. As a matter of proven fact that cloud will propose each storage and computation at low-value amounts makes it well-liked. This conjointly makes it a beneficial and powerful proposition for the long run. Despite showcasing tremendous potential by the cloud services, securing information within the cloud is the major cause of concern. These security issues conjointly build the employment of cloud services less versatile. During this analysis model work, we have a tendency to represent a secure service overloading Fusion Model that allows knowledge to be held on firmly in the cloud whereas at the same time permitting operations to be performed thereon with no compromise of the sensitive elements of the information.

Cloud has characteristics and associated behavior. However, its behavior might disagree in several things. The key of cloud overloading may be a cloud's services list that is additionally called the service overloading signature. It's the signature not the kind that permits cloud overloading (Figure 12.1).

If two clouds have the same variety and kind of services within the same order, they're the same to possess an equivalent signature, and if they're victimization distinct services name, it doesn't matter. However, they're distinct victimization networks.

FIGURE 12.1 Cloud behavior in overloading.

To overladen a cloud name, we ought to declare and outline all the clouds with an equivalent name but totally different signature individually.

The user ID and repair verify that the cloud ought to be invoked. A cloud referred to as initial matches the One-Time Password (OTP) obtainable service gives with the cloud decision and so calls the acceptable cloud execution; cloud overloading may be a logical technique of job severs as a cloud with totally different services perform essentially identical factors by an equivalent name.

Cloud overloading is a service name that provides the services with the same name having a totally different signature. To overload a service's name, all we'd like to try is to declare and outline all the functions with an equivalent name, however with totally different signatures, separately.

The previous analysis delineated many security problems impacting the cloud computing model and authentication. The safety state of cloud computing remains puzzled. It shifts computing perceptions entirely from the client viewpoint, outsourcing IT systems rather than adopting native solutions. Authentication is additionally being subject to changes. The static one-factor login is not any longer viable. Systems square measure is usually prone to information breaches, and attacker's square measure increases their arsenal of tools and techniques for brute-forcing secret hashes and bit by bit lowering trust of secret usage. An analysis on authentication is targeted on augmenting existing mechanisms or utilizing QR secret writing, Short-Message-Service (SMS) messages, OTPs, or life science–supported growing mobile technology for production of novel master's degree approaches, while reaching to attain necessary security properties like good forward secrecy and completeness. Nevertheless, very little attention is paid to the underlying infrastructure. Few have looked into harnessing the cloud computing technology to create authentication a lot of sturdy, in response to the adversity of web threats to public clouds, notably to the management interfaces.

Cloud may be a service field. The matter of indisputable fact that cloud will propose each storage and computation at low-value amounts makes it standard. This conjointly develops it as an awfully helpful and powerful proposition for the longer term. These security considerations conjointly create the employment of cloud services less versatile. During this analysis model work, we tend to represent a secure service overloading Fusion Model that lets information to be hold on firmly in cloud while at same time permitting operations to be performed thereon with no compromise of the sensitive components of the information.

12.4.1 Cloud Overloading

Cloud has characteristics and associated behavior. A bring service will see that is its behaviors. Still its behavior could dissent in several things. The key of cloud overloading could be a cloud's services list called the service overloading signature. It's the signature not the kind that permits cloud overloading. If two clouds are having the same variety and kind of services within the same order, they're said to possess an equivalent signature, and if they're mistreated distinct services name, it doesn't matter. Notwithstanding, they're distinct mistreatment networks.

To full a cloud name, the permit has to declare and outline all the cloud with the same name but numerous signatures individually.

Cloud overloading a service name provides the services with the same name having a totally different signature. To overload, a service's name, all we tend to need to try is to, declare & outline all the functions with a common name, however totally different signature, separately.

12.4.2 SERVICE OVERLOADING

Service overloading provides some way to outline and use services like upload, transfer, and user-defined kinds like cloud and security services. Let's develop an "overloaded" cloud, for example, the user a utility of services overloading.

Cloud overloading permits associates of its in-built service to be overladen in a precise cloud. Though solely those who have an intuitive that means for a specific cloud ought to really be overladen.

Cloud overloading is one of the foremost difficult and exciting characteristics of the cloud. Service that already exists within the cloud will solely be overladen. Overloading can't alter either the fundamental model of a service or its place within the order of precedence. Service overloading will be administrated by means of either a primarily based cloud or drive cloud.

If any clouds have multiple services with the same names but totally different parameters time (PT), they're aforementioned to be overladen. Service overloading permits users to use the same name for various activities to perform same or totally different service within the same cloud network.

12.5 THE PROPOSED MODEL

Management interfaces of cloud computing services may expose to a set of problems that have an effect on the web. They are conjointly generally accessed by many customers and might be scrutinized by attackers outside the cloud network for public clouds. As such, the model bestowed during this section aims to deal with the need of constructing authentication additional sturdy at the frontier of the cloud network.

This model explains service overloading that is employed to apply completely different service in one state of affairs. Cloud is enforced by mistreatment service overloading. The pure virtual and non-virtual networks may utilize in static and dynamic service. These services overloading technique is employed by base cloud. Service overloading is versatile because the multiple service is utilized in single side.

- Overloaded cloud offers the network the pliability to decide an analogous cloud for various forms of service.
- Service overloading is completed for service reusability, to save lots of efforts, and conjointly to save lots of price and supply safer network.

It is a plan of service-oriented network (SON). Service overloading may be a SON conception that permits datacenter to outline 2 or additional service with identical name and within the same scope. Every service includes a distinctive signature (or header) that springs from: service / service ID name (Figure 12.2).

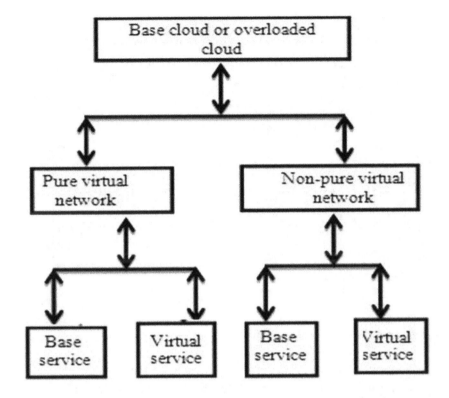

FIGURE 12.2 Service overloading architecture.

12.6 OVERLOADING AUTHENTICATION SYSTEM

The planned cloud overloading authentication technique is split into two tiers. The first tier, cloud overloading authentication, uses the encoding. A secret writing system has been used in traditional validation and authentication technique. The second tier, cloud overloading validation and authentication, necessitates the customer to add or insert alternative countersign from his respective hardware device like tablet or mobile that secures a unique id (such as IMEI in case of the mobile phone) which has been produced from the cloud server and directed toward the customers' hardware devices. Once the login window is visible through the computer's user interface system and the customer goes through the cloud id and password of the desired cloud, an auto system-generated OTP would be spawned from the server and directed toward the registered hardware device of the cloud user's. The user then enters their countersign from their personal hardware device and directs the message to the cloud server. The customer or user then might send the device an SMS straight forwardly or might use cloud mobile application like any customary service for specific cloud server. Then all the countersign of the standard users or customers go into the computer's login interface because the password entered within the customer's or user's individual device in interface is documented at the (cloud) end of the server. If any of the authentication fails, then the cloud access will be repudiated.

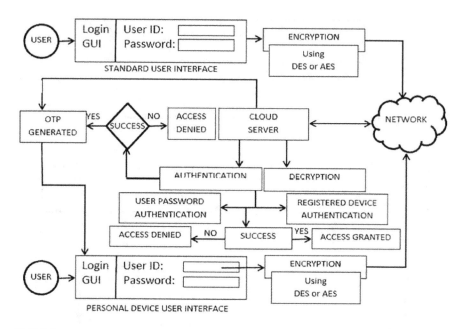

FIGURE 12.3 Functioning of the planned multi-authentication system.

Stepwise Functioning of the Proposed Framework Is Enlightened Below:
Stage 1: Customer or user feeds the URL by the supplier of cloud services in the website of his browser. The user then logs into the graphical user interface of the browser.

Stage 2: Customer or user feeds his tier1 identifications, that is, password and username. Later these authorization units were approved to the cloud server for authentication.

Stage 3: The cloud server verifies the authentication of credentials of tier 1 users. If the credentials like username and password are accurate, then the cloud server sends an OTP to the client of the particular cloud user.

Stage 4: After getting the OTP the user feeds the OTP in their registered individual device's interface and then sends it to the cloud.

Stage 5: The cloud server then verifies the password for authentication by the device's ID.

Stage 6: If authentication becomes successful, the server establishes a straight contact between customer and cloud server, and the user is permitted to access the cloud services (Figure 12.3).

12.7 IMPLEMENTATION AND RESULTS

Cloud overloading technique is predicated on the service overloading technique and therefore the implementation of labor is finished in one phase.

First the Fusion Model is prepared and then enforced on Citrix xen server and tested it for 12 months. Once satisfactory results area unit obtained with regard to the

Low Cost Robust Service Overloading Fusion Model 211

projected analysis work, then for supporting additional hardware and deep study the CloudSim machine and OMNeT++ are employed. The cloud sim simulation setting is made by composing four components; two overladen Cloud, and two shoppers, all the four parts are enforced on computers having P4 3.0 CPU, 2GB RAM. The service overloading technique is implemented equally on all clouds.

Finally, the results are complete that is mentioned within the following sections.

12.7.1 THE SIMULATION ENVIRONMENT

The subsequent tools and techniques used for the implementation of the projected technique are completed using:

1. CloudSim
2. OMNeT++
3. Jswings
4. Cloud overloading technique
5. Service overloading technique

Simulator	cloudsim-3.0.3&OMNeT++
Simulator Base	Java and C++

We implement our mechanism discrimination Java and C++language and CloudSim and OMNeT++ machine. The method is conducted on a pc with minimum of P4 3.0 central processing unit processor running at P4 3.0 GHz, 4 GB RAM.

It prices a lot of time if the users calculate the standard exponentiations while not outsourcing, and also the time value is increasing with the expansion of the bit length of exponent. By using our secure Fusion Model, the time value will be greatly reduced, and also the time value is sort of unchanged once the bit length of exponent increases.

12.7.2 SIMULATION RESULTS AND ANALYSIS

As displayed in Figure 12.4, the value of communication in SOFM might be around 1,526 bytes whereas in Authentication Protocol for Cloud Computing (APCC) was around 1,588 bytes. We wish to determine the actual fact that the info communication value of SOFM is 96 bytes (Table 12.1).

Figure 12.5 demonstrates that the time of authentication of SOFM is roughly 500 msw, whereas that of APCC is 514 ms, i.e., the time of verification SOFM is 97% of that of APCC.

After the verification the time result authorize that by simulation it has been analyzed that the communication cost is less of SOFM and result analyzed time is comparatively shorter (Table 12.2).

Calculated value of MS is 37.

Whereas, SAP value is 220 MS.

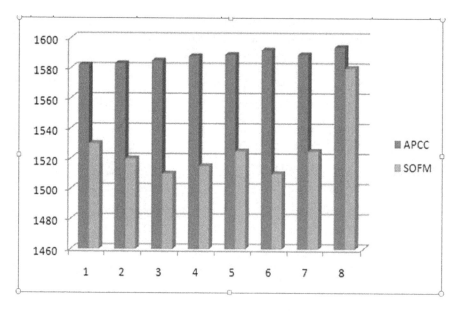

FIGURE 12.4 Comparison of communication cost.

TABLE 12.1
Comparison of Communication Cost

	Location	APCC	Location	SOFM
1	Lab	1582	Europe	1,530
2	Lab	1583	Arab States	1,520
3	Lab	1585	Africa	1,510
4	Lab	1588	Asia	1,515
5	Lab	1589	North America	1,525
6	Lab	1592	South/Latin America	1,510
7	Lab	1589	Arab States	1,525
8	Lab	1594	North America	1,580

That is, the calculated time of customer for APCC is 17% of SAP.

Figure 12.6 shows exemplifier calculation time of server for APCC as around 193 MS and for SAP as 276 MS. So we can deduce, the calculation time of server for APCC is 70% of that for SAP.

The result of simulation of both customer and server of APCC is lighter in weight than those of SAP (Table 12.3).

As shown in Figure 12.7, in APCC, the calculation of time is as follows:
Time of client is 37 MS.
Time of server is 192 MS.

From the calculation, it has been deduced that the customer's calculation time is 19% that of server in APCC.

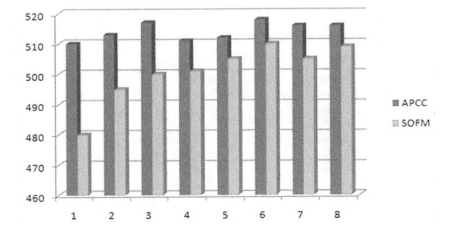

FIGURE 12.5 Comparison of authentication time.

TABLE 12.2
Comparison of Authentication Time

	Location	APCC	Location	SOFM
1	Lab	510	Europe	480
2	Lab	513	Arab States	495
3	Lab	517	Africa	500
4	Lab	511	Asia	501
5	Lab	512	North America	505
6	Lab	518	South/Latin America	510
7	Lab	516	Arab States	505
8	Lab	516	North America	509

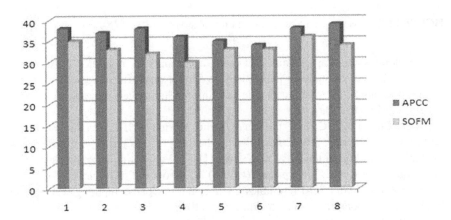

FIGURE 12.6 Assessment of computation time of client.

TABLE 12.3
Comparison of Computation Time of Client

	Location	APCC	Location	SOFM
1	Lab	38	Europe	35
2	Lab	37	Arab States	33
3	Lab	38	Africa	32
4	Lab	36	Asia	30
5	Lab	35	North America	33
6	Lab	34	South/Latin America	33
7	Lab	38	Arab States	36
8	Lab	39	North America	34

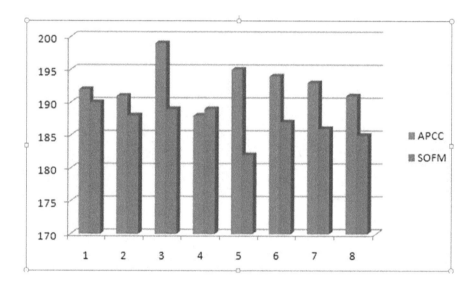

FIGURE 12.7 Assessment of computation time of server.

The cloud computing permits users with a platform for restricted presentation to outsource its computational jobs to certain further powerful servers. As a consequence, the more lightweight user side can link further servers and add to a bigger scalability (Table 12.4).

According to APCC, as established in the previous IBE and IBS systems, a safe APCC is planned. APCC is similar to the TLS protocol as both use RSA key exchange algorithm protocol as stated earlier [14]. APCC for cloud is a collection of three equivalent levels. The top-level (level 0) is the root of PKG. The level-1 is sub-PKGs. Every node in level-1 resembles a data center in cloud computing. The bottom level (level 2) is identified users in cloud computing.

In APCC, each node has an exclusive name. This method is applicable for different levels, but the proposed method (SOFM) encumbered lesser time for authentication

TABLE 12.4
Assessment of Computation Time of Server

	Location	APCC	Location	SOFM
1	Lab	192	Europe	190
2	Lab	191	Arab States	188
3	Lab	199	Africa	189
4	Lab	188	Asia	189
5	Lab	195	North America	182
6	Lab	194	South/Latin America	187
7	Lab	193	Arab States	186
8	Lab	191	North America	185

for the service provided or creating the environment and performs the job; hence, SOFM is better than APCC.

As we have already figured the assessment of communication rate among the two dissimilar techniques, it has been observed that only dominant communication is considered, i.e., certificate, signed or encrypted messages, which may have the greatest consumptions of the network bandwidth.

As shown in the result communication rate of IBHMCC is one IBS signature and one IBE cipher-text. Yet, in SOFM, the communication cost is simply cipher-text. With the assist of SOFM the IBS signature and communication rate is condensed. The assessment of communication rate among the two dissimilar techniques is revealed here. In this part, initially, we represented a novel Service Overloading Fusion Model for Cloud Computing (SOFM). Then, Cloud Overloading (CO) and Service Overloading (SO) techniques for cloud computing are projected. Finally, a SOFM is build based on the technique of overloading. Being certificate-free, SOFM aligns well along with the demands of cloud computing. Through simulation experiments, it is shown that the SOFM model is more lightweight and efficient than APCC. Here the implementation of the CO technique and its analysis is shown and proved. The SO technique can offer rapid and protected entree to cloud services below the cloud security situation even when consumers are in motion. When we compare it to the conventional SO technique, this authentication method does not need to take any device. There is no cost for purchasing or replacing all hardware, and it can offer simple access to the cloud service with suitable authentication procedures. Lastly, the proposed work has proved its superiority above SOFM and APCC methods.

12.8 CONCLUSION

Our model has been conceding to overcome safety glitches throughout the data transmissions over the cloud network. Our planned model offers the particular key management system to link security and cryptography for the aim of data integrity. The planned model has been clearly outperformed the present work compare. During this analysis work, the implementation of the CO Technique and its analysis is shown and proved. The SO technique will offer speedy and protected entry to cloud services underneath the cloud security atmosphere even once users are in motion. This authentication method doesn't need to hold any device perpetually compared to the traditional SO technique. There's no value for getting or buying any hardware and might offer a quick entree to the cloud service with a suitable authentication method. Lastly, our research work has proved its superiority over SOFM and APCC techniques.

REFERENCES

1. Andersen T.J. (2001). Information technology, strategic decision-making approaches and organizational performance in different industrial settings. *Journal of Strategic Information Systems*, 10, 101–119.
2. Kagal, L. et al., (2001). Trust-based security in pervasive computing environments. *Computer*, 34, 154–157.
3. Patnayakuni N. and Rai A. (2002). Towards a theoretical framework of digital supply chain integration. *European Conference on Information Systems (ECIS)*, Gdańsk, June 2002. [Online]. Available: http://is2.lse.ac.uk/asp/aspecis/20020127.pdf.
4. Rico, D. (2005). Practical metrics and models for return on investment. *TickIT International Journal*, 7(2), 10–16.
5. Santos, N. et al., (2009). Towards trusted cloud computing. *HotCloud*, 9(9), 3.
6. Tadapaneni, N.R. (2020). Cloud computing security challenges. *International Journal of Innovations in Engineering Research and Technology*.
7. Turab, N.M., Taleb, A.A. and Masadeh, S.R. (2013). Cloud computing challenges and solutions. *International Journal of Computer Networks & Communications*, 5, 209–216. Doi: 10.5121/ijcnc.2013.5515.
8. Itani, W. et al., (2009). Privacy as a service: privacy-aware data storage and processing in cloud computing architectures. In: *2009 Eighth IEEE International Conference on Dependable, Autonomic and Secure Computing*, 711–716.
9. Mather, T. Kumaraswamy, S., and Latif, S. (2009). *Cloud Security and Privacy: An Enterprise Perspective on Risks and Compliance*. Oreilly& Associates Inc, 2009.
10. ENISA. (2009, Nov.). Cloud computing: Benefits, risks and recommendations for information security [Online]. Available: http://www.enisa.europa.eu/act/rm/files/deliverables/cloud-computing-riskassessment.
11. Gritzalis, D. (2004). Embedding privacy in IT applications development. *Information Management & Computer Security*, 12(1), 8–26.
12. Katz F.H. (2005). The effect of a university information security survey on instructing methods in information security. In: *Proceedings of the Second Annual Conference on Information Security Curriculum Development*, Association for Computing Machinery, New York, 43–8.
13. Tang C.S. (2006). Robust strategies for mitigating supply chain disruptions. *International Journal of Logistics: Research and Applications*, 9(1), 33–45.
14. Kanhere, V. (2009). Driving value from information security: a governance perspective. *ISACA Journal*, 2.
15. Vishwakarma, R.P., Tamrakar, S., and Sharma, R.K. (2017). Data security and service overloading in cloud computing –an overview. *International Journal of Computer Sciences and Engineering*, 5(8), 177–180.
16. Garfinkel, T. et al., (2003). Terra: a virtual machine-based platform for trusted computing. *ACM SIGOPS Operating Systems Review*, 37, 193–206.
17. Chen I.J. and Paulraj, A. (2004). Towards a theory of supply chain management: the constructs and measurements. *Journal of Operations Management*, 22, 119–150.

13 Load Balancing Based on Estimated Finish Time of Services

Rajesh Sachdeva
Dev Samaj College for Women, Firozpur

CONTENTS

13.1 Introduction 217
     13.1.1 Sorts of Cloud Computing 218
13.2 Load Balancing... 218
13.3 Related Work 219
13.4 Proposed Load Balancing Algorithm... 219
13.5 Proposed Methodology... 221
13.6 Imitation and Outcome Analysis 223
     13.6.1 CloudSim 223
     13.6.2 Netbeans (Software)... 223
13.7 Experimental Results.. 224
     13.7.1 Waiting Time of Proposed Algorithm 225
     13.7.2 Turnaround Time of Proposed Algorithm 225
     13.7.3 Processing Cost of Proposed Algorithm 226
13.8 Conclusion and Future Work 226
References... 226

13.1 INTRODUCTION

Cloud computing is an Internet grounded facility benefactor in which operatives are permitted to access facilities only when required. By using concepts of parallel computing, grid computing and distributed services, Cloud computing allows dynamic allocation of resources [1]. Cloud computing has always been in focus because of its flexibility which results in improved facilities to end-users, the research community and especially software developers, and of its pliability capability on resource alliance. According to the latest trends in computing, everyone is shifting to the Cloud for their personal use or for their business because of its facilities and features. During the COVID-19 pandemic, its service providers have increased enormously as it provides facilities at a very low rate.

    The motto of this technology is to shift the computational facilities from local-based storage to the net that is shifting automation facilities offered and data off-site to a peripheral site. Cloud computing prototype is often called "pay-per-use model" because we have to pay only that we are using. We don't have to pay all time for it [2].

Virtualization is a key concept in Cloud computing. In virtualization, a machine (system) can be thought of as a number of virtual machines (VMs) or systems that are actually not physically present [3]. When a request is received from a client, the procedure fulfills the client's need by deciding which consumer will use the VM and which VMs are already assigned to some other consumer.

Load balancing is handled by using virtualization technology dynamically, which allows dynamic allocation of VMs and physical resources according to the load received from the client. Due to these advantages, it is mostly used by Cloud computing. Handling of different challenges like increase in VMs if required, output, availability, assigning VMs on-demand and fault tolerance in the load balancing is a tough task. Load balancing is basically a process in which the load is distributed among various clients or nodes to do an equal amount of labor.

13.1.1 Sorts of Cloud Computing

The following are available types of services in the Cloud:

- **Infrastructure as a Service:** This facility deals with the infrastructure part of cloud computing, i.e. if some client needs hardware through the Internet. Consumers can buy infrastructure facilities like servers or networks. The consumers pay only for the period they are using the facility, not all the time, i.e. clients need not buy infrastructure; they have to pay only for time services being used by the client [4].
- **Platform as a Service:** This service delivers simulation for run time, i.e. virtual atmosphere where one can build or test an application.
- **Software as a Service:** It delivers the consumer different package applications as a facility from many cloud benefactors through the Internet [5]. Users can use different types of software when needed without paying too much for them.

13.2 LOAD BALANCING

Cloud computing is ace of the broadest espoused and employed expertise in various groups. Due to the pliable architecture of the Cloud, many groups are setting their own Cloud, which ultimately increased the number of users.

Reliability is one of the foremost apprehensions in the Cloud as to which algorithm is more reliable for what type of architecture [6]. Cloud computing is being used by different groups such as social networking websites and online applications design by Google doc, and several clouds are also used for online software testing [7,8].

To improve utilization of resources, improve overall performance and to properly assign/reassign computing resources to the clients and save energy, load balancing has been emphasized in the Cloud Computing atmosphere [9,10].

13.3 RELATED WORK

In the existing load balancing algorithm, a methodology is used in which data center broker (DCB) maintains an index table. The index table contains three entries: *(i) number of VMs, (ii) VM status, and (iii) current load* on VM [11,12]. When client requests the DCB, it will check entries in the index table and compute the time with each Simulated Engine or VM [13,14]. The VM having the least completion period will be designated from the available VM's. The index table will be updated with ID and the corresponding status of the currently selected VM based on SS or TS [15].

But in a real environment, there are many VMs available. When the client requests a job, it is asked for DCB. Then, DCB calculates execution time with all the VMs. This methodology results in spending/wasting too much time in finding the relevant VM to run a job requested by the user, which may result in wastage or underutilization of resources. The problem is similar to linear search, where it has to compare with every element and ultimately increase reply time and computation time of the task.

We got many research gaps in the cost scheduling algorithm, which are listed below:

1. When we run program, execution time is computed for all the available VMs.
2. This practice wastes excessively much spell in the verdict of significant VM.
3. There can be underutilization of resources in the cloud environment.

13.4 PROPOSED LOAD BALANCING ALGORITHM

In our proposed algorithm, we brood over two factors: one is the actual immediate processing power of VM and the second is the size of allocated work, and workload is alienated across the VM by considering three cases and considering three factors:

A. Completion time of cloudlet;
B. Availability of virtual machine;
C. Current load of the virtual machine.

In order to estimate the end of service time of VMs, we consider two factors. The selection of VM for the next job depends on the three cases which are as follows:

1. **CASE 1:** Include all the possible VMs in the candidate set (used for selecting VM) for computing the completion time of job. This case is applicable If all the VMs are Available for use.
2. **CASE 2:** Consider only those VMs that are free if some VMs are doing work and some are free.
3. **CASE 3:** If all VMs are Busy in doing computational work, then completion time is checked with every VM and then added to the candidate set for further process.

Based on these three cases, their corresponding result will be:

1. If a job/cloudlet belongs to **Case 1**, then we consider all VMs and completion time is computed with all VMs.
2. If a job/cloudlet belongs to **Case 2**, then we consider only those VMs which are available and ignore VMs which are busy.
3. If a job/cloudlet belongs to **Case 3**, then firstly, we sort them in descending order in terms of processing power, i.e. most powerful VM will be assigned job in that case.

It is very difficult to access the immediate processing power of VMs. It may depend on which policy is being used in Cloud computing at two levels. Firstly, we will try to scheme End time [4] and then we try to complete it with the proposed formulas, which then can be used to compute the processing power of simulated core depending on various arrangements [16,17].

The main goal of our projected algorithm is to progress the routine of the structure by improving the response time, waiting time, turnaround time and processing cost in these four arrangements:

1. host-level reflecting SS and in VM level reflecting SS;
2. host-level reflecting SS and in VM level reflecting TS;
3. host-level reflecting TS and in VM level reflecting SS;
4. host-level reflecting TS and in VM level reflecting TS.

To determine the response time the following formula can be used:

$$\text{Response Time} = \text{End Time} - \text{Arrival Time}$$

The End time can be determined by the following method:

$$\text{End Time} = \text{Total number of instruction} / (\text{Capacity} \times \text{Cores})$$

To calculate time for implementation the following method can be used:

$$\text{Execution Time} = \text{Total number of instruction} / (\text{Capacity} \times \text{Cores})$$

Capacity is the mediocre handling capacity of a core for the work, and core is the number of computation rudiments.

A Capacity parameter is used to measure the performance of processing a job on VM which depends on the type of arrangements policy for host and VM. The total processing capacity of a host is measured by the total number of its core and corresponding processing power. However, In a real-time environment each consumer requires certain cores and facilities shared by numerous consumers concurrently. If the total amount of the core is superior to the total amount of bodily core, then we use the idea of virtual core, and in that case, the processing power of each core that is virtual can be lesser than of bodily core. Formulas of calculating the Capacity is mentioned above. There are two stages of arrangements: scheduling VMs to share physical host resources and scheduling jobs to share VM resources. Our projected load balancing procedure that can be used to compute Capacity is as follow:

1. host-level reflecting SS and in VM level reflecting SS:

$$\text{Capacity} = \sum_{i=1}^{n} \left(\frac{\text{Processing power of core } i}{\text{no of cores in the host}} \right) \qquad (13.1)$$

2. host-level reflecting TS and in VM level reflecting SS:

$$\text{Capacity} = \frac{\sum_{i=1}^{np} \text{Processing power of core } i}{\text{Max}\left(\sum_{j=1}^{\alpha} \text{cores}(j), np \right)} \qquad (13.2)$$

where cores (j) is the number of cores that job j needs, α is the total job in VM that contains the job.

3. host-level reflecting SS and in VM level reflecting TS:

$$\text{Capacity} = \frac{\sum_{i=1}^{np} \text{Cap}(i)}{\text{Max}\left(\sum_{k=1}^{\beta} \sum_{k=0}^{\gamma} \text{cores}(g), np \right)} \qquad (13.3)$$

where β is the number of VMs in the current host. Υ is the number of jobs running simultaneously in VMk.

4. host-level reflecting TS and in VM level reflecting TS:

$$\text{Capacity} = \frac{\sum_{i=1}^{np} \text{Cap}(i)}{\text{Max}\left(\sum_{j=1}^{\delta} \text{cores}(j), np \right)} \qquad (13.4)$$

where δ is the total job of the considered host (Figure 13.1).

13.5 PROPOSED METHODOLOGY

Step 1: Firstly, we have to Create a DCB and status table that will maintain subscript or index of the VM and VM that are doing work during that duration and check whether it has completed its task. Initially, no work is assigned to any machine.

Step 2: Below are the instances that have to be considered during time estimation of the jobs by DCB that will check status index table when it receives a request from consumer.

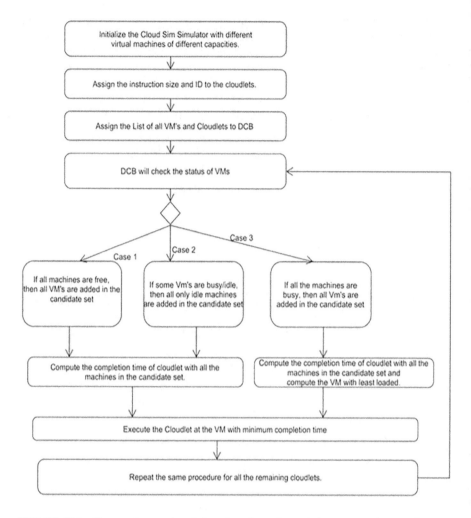

FIGURE 13.1 Proposed procedure is based on the estimated time to complete service.

Instance 1: If altogether VMs are free to use, consider all of them and add them in set to calculate their time to complete.

Instance 2: If some VMs are doing work and others are free to use, consider only those that are free and consider them in set for calculation.

Instance 3: If in case all VMs are busy, then consider all for calculation and consider them in set.

Step 3: A VM will be chosen from the set to complete the request of consumer who requested DCB to complete its job, and the following instances will be carried out further.

If Instance 1 is chosen, As All the VMs are free and machine having minimum completion time will be chosen.

Load Balancing by Estimated Finish Time of Services

If Instance 2 is chosen, Some VMs are free to use while others are processing their previous jobs. So consider only those machines that are free to use and consider machine with minimum time.

If Instance 3 is chosen, All the VMs are busy doing their previous work. So, we have to run cloudlets and check the minimum time with each machine and the machine having minimum time will be chosen.

Step 4: VM that is selected by the methodology provides the ID of VM for further execution.

Step 5: Jobs that the VM identifies are assigned by DCB.

Step 6: DCB informs the proposed methodology about the novel distribution.

Step 7: Current status of our i-table of VM and that of job will be updated.

Step 8: When the VM completes processing necessities of the job and DCB is responded about job, status i-table will be updated and one job is reduced from the i-table.

Step 9: Go to step 2

13.6 IMITATION AND OUTCOME ANALYSIS

The Imitation and Outcome Investigation will be completed by means of the CloudSim.

13.6.1 CLOUDSIM

CloudSim provides an imitation environment where we can test the logic of the proposed algorithm. CloudSim basically includes all the computing infrastructures of Cloud a services. Without knowing the actual infrastructure, users can test their logic's performance in a real-time environment.

For analysis of different load balancing plans, parameters for deploying applications DCB configuration and user base are configured.

Concentrating on logic without dealing with low level of structure related to CloudSim, many programmers and researchers are using CloudSim.

The key feature of CloudSim is to support the imitation of bulky scale Cloud computing atmosphere comprising data centers, and its advantages include effectiveness and elasticity. CloudSim helps to evaluate the hypothesis in the environment so that one can find results of imitation in a controlled environment.

13.6.2 NETBEANS (SOFTWARE)

NetBeans is an integrated development environment (IDE), i.e. environment that provides development (using drag and drop) and is used for mounting with programming languages like Java, PHP, C/C++ and HTML5 [18]. NetBeans is inscribed in Java language and can route on Operating System like Linux, Microsoft Windows and different podiums supporting Java VMs. Third-party Software or Applications can also extend the NetBeans Platform for using [2] (Figure 13.2).

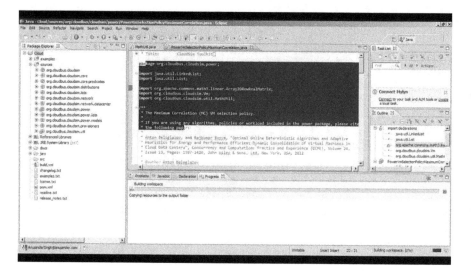

FIGURE 13.2 CloudSim environment in net beans.

13.7 EXPERIMENTAL RESULTS

See Table 13.1.

TABLE 13.1
Showing the Different Parameters of the Proposed Methodology by Running a Different Number of Cloudlets

S. no.	No of Cloudlets	Average Turnaround Time	Total Execution Time	Average Waiting Time	Total Processing Cost
1	5	0.41	1.24	0.16	3.80
2	100	5.50	31.43	5.19	95.91
3	200	10.80	62.92	10.49	191.99
4	300	16.20	94.76	15.89	289.13
5	400	21.47	126.17	21.15	384.96
6	500	26.74	157.66	26.43	480.87

13.7.1 WAITING TIME OF PROPOSED ALGORITHM

See Figure 13.3.

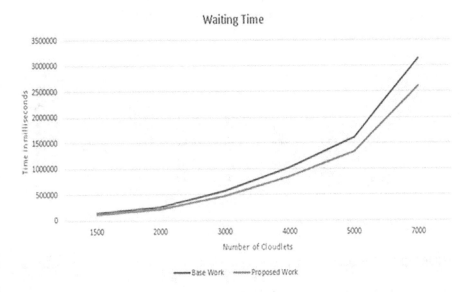

FIGURE 13.3 Waiting time of the proposed methodology.

13.7.2 TURNAROUND TIME OF PROPOSED ALGORITHM

See Figure 13.4.

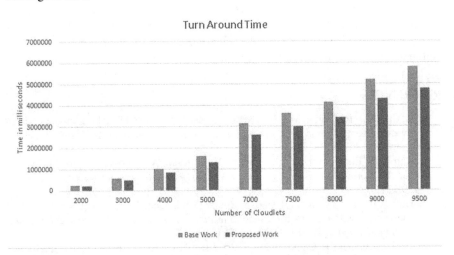

FIGURE 13.4 Turnaround time of the proposed procedure.

13.7.3 Processing Cost of Proposed Algorithm

See Figure 13.5.

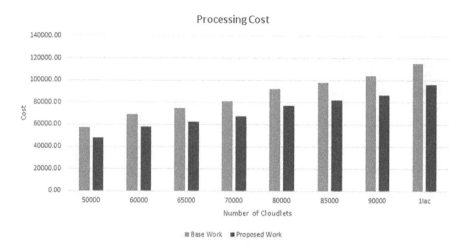

FIGURE 13.5 Processing cost of proposed methodology.

13.8 CONCLUSION AND FUTURE WORK

From the imitation results of our algorithm, we can say that there is an improvement in performance in some fields as compared to the previous one. Average response time is decreased, and if we consider the total processing cost of VM in Million Instructions Per Second (MIPS), Cost also decreases.

As the number of cloudlets increases, the proposed model performs much better. The forthcoming work involves a dynamic shifting of load from one VM to another VM in case of sudden failure. There must be an obligation to handle such a condition. Automatic fault tolerance may be added in forthcoming work.

REFERENCES

1. Ajith, S.N., Hemalatha, M., An approach on semi-distributed load balancing algorithm for cloud computing system, *International Journal of Computer Applications*, 56(12), 1–4, 2012.
2. Caron, E., Rodero-Merino, L., Desprez, F., Muresan, A. Auto-scaling, load balancing and monitoring. In *Commercial and Open-Source Clouds* Research Report, 27, January 2012.
3. Hiranwal, S. Roy, K.C. Adaptive Round Robin scheduling using shortest burst approach based on smart time slice. *International Journal of Computer Science and Communication*, 2(2), 319–323, July-December 2011.
4. Wickremasinghe, B. Cloud analyst: a cloud-sim-based tool for modeling and analysis of large scale cloud computing environments. MEDC Project, Report 2010.
5. Buyya, R., Ranjan, R., Calheiros, R. N. Modeling and simulation of scalable cloud computing environments and the cloudsim toolkit: challenges and opportunities. In *2009 International Conference on High Performance Computing & Simulation*, IEEE Computer Society, IEEE, Leipzig, 1–11, June 2009. doi: 10.1109/HPCSIM.2009.5192685

6. Anthony T., Velte, T.J., Elsenpeter, R. *Cloud Computing a Practical Approach*, Tata Mcgraw-Hill Edition, 2010.
7. Sun Microsystems, Inc. *Introduction to Cloud Computing Architecture* Whitepaper, 1st Edition, June 2009.
8. Howell, F., Macnab, R. SimJava: a discrete event simulation library for Java. *Simulation Series*, 30, 51–56, Jan. 2008.
9. Buyya, R., Murshed, M. GridSim: a toolkit for the modeling and simulation of distributed resource management and scheduling for Grid computing. *Concurrency and Computation: Practice and Experience*, 14, 1175–1220, Nov. 2002.
10. Armbrust, M., Fox, A., Griffith, R., Joseph, A.D., Katz, R., Konwinski, A., Lee, G., Patterson, D., Rabkin, A., Stoica, I., Zaharia, M. *Above the Clouds: A Berkeley View of Cloud Computing*. EECS Department, University of California, Berkeley, CA, Technical Report No., Ucb/Eecs-2009-28, 1–23, February 2009.
11. Lee, R., B. Jeng Load balancing tactics in cloud. In *International Conference on Cyber Enabled Distributed Computing and Knowledge Discovery*, IEEE, Beijing, 447–454, 2011.
12. Endo, P.T., G.E. Gonçalves, Kelner, J. A survey on open-source cloud computing solutions.
13. Kalagiakos, P. Karampelas, P. Cloud computing learning. In *2011 5th international conference on application of information and communication technologies (AICT)*, IEEE, Baku, 1–4, Oct. 2011.
14. Yadav, A.K., Bharti, R., Raw, R.S. SA^2-*MCD*: secured architecture for allocation of virtual machine in multitenant cloud databases. *Big Data Research: An International Journal*, Elsevier, 24, 2021, Doi: 10.1016/j.bdr.2021.100187.
15. Yadav, A.K., Bharti, R., Raw, R.S. Security solution to prevent data leakage over multitenant cloud infrastructure. *International Journal of Pure and Applied Mathematics*, 118(07), 269–276, 2018.
16. Kumar, M., Yadav, A.K., Raw, R.S. Global host allocation policy for virtual machine in cloud computing. *International Journal of Information Technology*, Springer, 1–9, 2018.
17. Dixit, A., Yadav, A.K., Raw, R.S. A comparative analysis of load balancing techniques in cloud computing. In *International Conference on Communication and Computing Systems (ICCCS-2016)*, 2016, India, 1031–1036, Doi: 10.1201/9781315364094-67.
18. Mishra, R., Jaiswal, A.P. Ant colony optimization: A solution of load balancing in cloud. *International Journal of Web & Semantic Technology*, 3(2), P33, April 2012.

14 Blockchain-Enabled Smart Contract Optimization for Healthcare Monitoring Systems

Nitima Malsa and Vaibhav Vyas
Banasthali Vidyapith

Pooja Singh
Amity University

CONTENTS

14.1 Introduction 230
    14.1.1 Blockchain 230
    14.1.2 Ethereum...232
    14.1.3 Smart Contract..233
        14.1.3.1 Smart Legal Contracts 234
        14.1.3.2 Decentralized Autonomous Organization (DAO)..............234
        14.1.3.3 Application Logic Contracts (ALCs)................................234
    14.1.4 Gas Optimization Techniques 236
14.2 Literature Survey 236
14.3 Methodology...239
    14.3.1 Designing of Healthcare System Smart Contract on Ethereum Blockchain..239
    14.3.2 Writing Healthcare System Smart Contract Using Sublime Text 3 and Remix Ethereum IDE 239
    14.3.3 Optimization Techniques Applied on Smart Contract to Reduce the Gas Cost..245
14.4 Results...245
14.5 Conclusion 248
References 249

14.1 INTRODUCTION

This chapter is organized as follows: Section 14.1 briefly introduces blockchain, Ethereum, and smart contract with their work process in the health system. Section 14.2 presents a literature review about the existing Healthcare systems using blockchain and various optimization techniques available for optimizing smart contracts. Section 14.3 describes the methodology with implementation details consisting of mainly three steps. Section 14.4 presents the results of the work. Finally, Section 14.5 concludes the chapter. The last section includes useful references.

Blockchain emphasizes decentralized data storage. It is a chain of blocks also known as a distributed ledger. It is extensively applicable in all areas where the stakes play a significant role; therefore, there must be a state of confidence between the communicating parties placed at distributed locations. Blockchain is an emerging technology and has been proved in various fields like healthcare, financial, insurance, the Internet of Things (IoT), etc. Bitcoin is the first cryptocurrency based on blockchain technology [1]. A smart contract is an efficient tool that is used in healthcare systems to manage the system with trust and transparency. If someone tries to manipulate the data, it is not possible as smart contracts are immutable. Ethereum is a blockchain-based public platform. It supports improved and customized smart contracts [2].

The existing healthcare system suffers from performance constrictions as they are not transparent, and sometimes patients get wrong treatment and reports are exchanged. To remove these types of problems, blockchain-based applications can be built. The blockchain-based implementation makes the process faster, transparent, and tamper-proof.

This work addresses smart contract optimization to a blockchain-based solution using smart contracts. The main purpose of using blockchain in the Healthcare system is to speed up the process. The chapter discusses various optimization techniques, some of which are applied in work to reduce the overall costs of the contract. Phases concerning the development of the smart contract for healthcare systems are as follows:

1. Designing of 'Healthcare system' smart contract on Ethereum blockchain.
2. Writing 'Healthcare system' smart contract using Sublime Text 3 and Remix Ethereum IDE.
3. Optimization techniques are applied on smart contracts to reduce the gas cost.

14.1.1 BLOCKCHAIN

Blockchain is a shared distributed ledger, i.e., a collection of records that are linked with each other. These records are strongly resistant to alteration and are protected using cryptography. A blockchain is essentially a digital ledger of transactions distributed across the entire network of computer systems on the blockchain and the data is stored the same as in the complete network, i.e., duplicated. The data is stored in batches called blocks that are linked together chronologically to form a chain. Each block contains several transactions; whenever a new transaction takes place, a copy of the record of that transaction is updated on every participant's ledger. Blockchain helps facilitate the records of transaction and asset tracking of anything that holds a value and allows all parties to monitor and analyze the status of an asset in real time.

Let us take the example of Google Docs. When a document is created and shared among a group of people, the changes that are made in that document are updated on everybody's end. The complete document is shared with the people instead of a copy or being transferred, this makes the document distributed. A decentralized distributed chain is formed in which anybody can access the document at the same time. All the modifications taking place in the document are recorded in real time, thus making everything transparent.

People might get confused between database and blockchain; the key difference between the two is that the data is stored into tables in a database, whereas in blockchain, the data is stored into blocks that are chained together. So, all blockchains can be databases but vice versa might not hold. Blockchain helps in reducing risks and frauds and makes the complete process more transparent. Figure 14.1 represents how blocks are connected in a blockchain.

Blockchain uses cryptographic keys to secure the identities of the individuals joining the network, and hash function is used to make the block unalterable.

Let us understand the process of a blockchain, as in blockchain, each block consists of these basic elements: the data stored inside the block, a nonce which is a 32-bit value used for generating a hash, the hash value, which is a 256-bit number, and the hash of the previous block which is used for linking.

The first block of the blockchain is called the 'Genesis Block'. On the creation of the genesis block, the hash value is generated with the help of nonce. The stored data inside the block is immutable and endlessly tied to the nonce unless it is mined. Hash values are used to prevent tampering, but due to the availability of high-speed computers, any attacker can recalculate the hash value and make the blockchain valid again after tampering. To prevent this, proof of work is needed. When a transaction is agreed upon between the parties, it needs to be approved before being added to the blockchain. The decision to add a transaction to a blockchain depends on the fact that the majority of the nodes should agree to call that transaction a valid one. This is done with the help of miners. The main task of a miner is to validate any unverified transaction and add them to the block by solving complex mathematical puzzles based on the hash function. This process is called proof of work. Miners find a nonce value that lies within the target requirement. For this, they need computers with high processing power. Solving the problem is called mining and the first miner to solve the puzzle gets rewarded in cryptocurrency. As the blockchain starts to grow, the

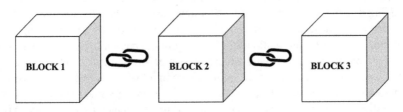

FIGURE 14.1 Blockchain.

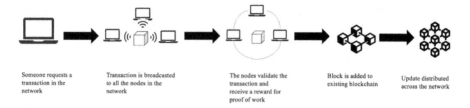

FIGURE 14.2 Blockchain transaction process.

problems start becoming more and more complex, thus making it harder for attackers to sabotage the chain. At any point in time, if somebody tries to enter a blockchain network, a complete copy of the blockchain is provided to that person. Each node is a computer connected to that network. Figure 14.2 shows the entire process of how transactions take place in a blockchain.

Blockchain has wide applications in sectors like marketing, finance, IoT, health, and insurance. Registration of properties using smart contracts and allowing insurers to check the transaction history are a few ways where blockchain is used in insurance [3]. Insurance companies use blockchain because it is immutable. The customer cannot falsely claim the insurance. The insurance company can verify the data remotely before approving the claim, thus making the whole process much more transparent. In healthcare insurance, the transactions will be the health records of the patient [4].

14.1.2 Ethereum

Vitalik Buterin, a Russian-Canadian and a developer, in 2013, published a white paper that proposed a platform that combined blockchain functionality in addition to a computer code being executed. Thus, Ethereum was born. Ethereum blockchain helps in creating programs that can make communication take place on the blockchain. Tokens are created in ethereum; these tokens represent digital assets of various kinds, track the ownership and execute its functionality according to a set of instructions written in the form of a program. Ethereum uses ether as a cryptocurrency. The transactional fees and processing services are paid by ether.

According to Ethereum, it can be used to "codify, decentralize, secure, and trade just about anything". Ethereum-based smart contracts can be compiled and deployed on Ethereum Virtual Machine (EVM). It is a runtime environment and is highly secure. There are two types of blockchains: permissionless and permissioned. Permissionless blockchains are those that require no permission to join the network. Anybody can join the network and interact with it. These blockchains are ideal for running and managing digital currencies and are also known as public blockchains. Permissioned blockchains are those where not everyone can join the network. The network administrator needs permission to allow a person to join the network. That is why it is also known as the private blockchain. Ethereum is a permissionless blockchain wherein anybody can join the network and communicate in it [5]. The key difference between bitcoin and ethereum is that bitcoin is used as a store of value, whereas ethereum decentralizes applications and services into complex financial agreements. After a person gets a few ethers, he/she needs a place to store them; for this purpose ethereum wallets are used. These wallets are apps that can be accessed from smartphones or laptops. These digital wallets are used for storing digital money such as ether or bitcoin. Users' private keys are stored inside the ethereum wallet which are later on used for accessing the ether stored. The owners of these private keys can only use them to spend the money stored in a wallet. Ethereum blockchain works similarly to bitcoin's blockchain; the only difference is that ethereum blockchain stores more information than transaction details, such as the details about the state or the present information, along with the user's balance, the smart contract codes, storage location, and details about the changes made [6]. Each node stores the following things:

Accounts: Details about the number of either each user has in their account.
Smart Contract Code: As smart contracts are stored in ethereum, it stores the rules that need to be verified before the transfer of money takes place.
Smart Contract State: Describes the state of the smart contracts.

Whenever some action is performed, all the nodes connected in the network validate that change and come to an agreement that that change took place. Bitcoin uses proof of work to keep the data in sync. Ethereum, on the other hand, does not use proof of work; it uses a different algorithm called proof of stake, which consumes less energy and is more secure. According to proof of stake, a miner can validate or mine a block according to the number of coins they hold. The more the number of coins a miner holds, the more mining power they'll have. Proof of stake requires a huge amount of energy as compared to proof of work and is less risky in terms of attackers sabotaging the network.

14.1.3 Smart Contract

A smart contract is a computer program that facilitates the exchange of money, property, shares, etc. It is a contract in which the terms and conditions between the two parties are written in lines of code. It is an executable code that runs on top of a blockchain to enforce and facilitate an agreement between two distrustful parties having without involving any trusted third party [7]. These smart contracts are distributed over the blockchain network along with all the code and agreements contained therein. It allows trusted transactions and agreements to take place between the stakeholders of the contract without the need for any central entity of third-party involvement. Thus, the transactions become more transparent, and as it is on the blockchain, it is irreversible too; hence nobody can alter any information in the smart contract once it is updated on the blockchain. Smart contracts are written in solidity which is a high-level object-oriented programming language. Smart contracts give transactions traceable, irreversible, and transparent. They are a part of any cryptocurrency. Solidity verifies and enforces the constraints at compile-time as opposed to runtime [6]. In some smart contracts, the logic and data can be separated. The logic of a contract can be changed or developed using a substitute contract while keeping in mind all the related conditions in the data. A few variables are already defined in solidity such as msg, tx, and block. These variables are available globally and the information can be accessed to a transaction invocation on the blockchain.

For example, the variables agree on the fetching of the source address, the amount of the Ether, and the information sent with the invocation transaction [6].

The key difference between traditional contracts and smart contracts is that the terms and conditions are put down on paper in traditional contracts. In contrast, in smart contracts they are executed by a coded program. Smart contracts are more secure than traditional ones; as in smart contracts no changes can be made once the program is executed, but in a traditional contract, the terms and conditions can be tampered with. Along with this, no third party is involved in a smart contract, but a third party is involved in cross verification of the terms and conditions in the traditional method. The main advantage of a smart contract over a traditional contract is that the smart contract is more efficient and maintains the contracting parties' history [8]. Smart contracts are of various types.

14.1.3.1 Smart Legal Contracts
These contracts involve strict legal recourses in case parties involved in the same were not to fulfill their end of the bargain.

14.1.3.2 Decentralized Autonomous Organization (DAO)
It is a theoretical company that runs by a code instead of people. DAO makes the hierarchy within a company less structured and can eliminate the middlemen within the company. Smart contracts enable DAO to engage in business work collaborations [9].

14.1.3.3 Application Logic Contracts (ALCs)
ALCs contain codes related to the application that works along with other smart contracts and programs on the blockchain. They help in communicating between devices and validating that communication (while in the domain of IoT).

A smart contract can be written by anybody and deployed to the network. The code written should be in solidity and the user should have enough ethers (ETH) to deploy the contract. As deploying a contract is a transaction, gas is paid in the same way gas is paid for a simple ETH transfer.

Let us understand the process of a smart contract with an example, suppose two individuals A and B are debating about a cricket match. Both the individuals have conflicting opinions about which team will win the match as both of them are supporting different teams. Both of them need to go elsewhere and won't be able to watch the match ending, so they decide to bet $100. A bet that team X will win, if that is the scenario, then B will have to pay $100 to A. B accepts the bet and adheres to some terms and conditions. But, neither of the two parties trusts each other to honor the best and there's no time to appoint a third party to oversee the same. Let us assume that both A and B are using a smart contract platform to automatically settle the bet. In this scenario, both A and B will link their blockchain-based identities to the contract and set the terms, so that the process is transparent as whosoever wins the match, the money will automatically be credited to his account from the loser's account. As soon as the match gets over, the program will surf the internet for the prescribed sources and identify the winning team. After identifying the winning team it relates to the terms of the contract; in case team X wins the game A will get money from B, and after intimating both parties $100 is transferred from B's account to A's account. Once the smart contract is executed, it will terminate and be inactive for all the time to come unless mentioned otherwise. All smart contracts revolve around the same principle, with the code being executed on predefined parameters and producing expected outputs only. A well-written smart contract should describe all possible outcomes of the contract. It is deterministic in nature, i.e., the same input will produce the same output [10].

So, the basic workflow of the smart contract includes identifying an agreement, setting the conditions, writing a computer program in such a way that the arrangement will be automatically performed when the parameters are valid, encrypting for security, executing the code when consensus is reached, and writing on the blockchain, the code is executed and outcomes are verified, in the last step all the computers in the network update their ledger so that new state is reflected. When there is a condition to which certain legal consequences are attached, the smart contract executes the corresponding statements and any potential contractual consequences [11]. A few benefits of the smart contract are the following:

1. It is faster than the traditional method.
2. Removes trust issues between the parties.
3. Eliminates the need for third party.
4. Can securely handle sensitive data.

The associated data will always remain in the blockchain platform it was executed on for future references.

Smart contracts can be used in health care in various ways. It can allow the patients to buy healthcare insurance policies using smart contracts and the details of the patient will be stored on the blockchain. This information can be shared with hospitals where the patient wants to get his/her treatment done. The sharing of the information will only be allowed if the patient has given permission. Similarly, the information regarding the treatment that the patient undergoes, the medicines prescribed by the consultant, and the bill of the treatment will be stored on the blockchain and further be shared by the hospital and pharmaceutical store with the insurance company. Once the complete data is stored on the blockchain, the insurance company can generate the insurance claim and the process can be automated by using smart contracts.

As in present times, patients obtain medical assistance from more than one institution. So, a patient looking for medical assistance from other hospitals needs to collect the data from various institutions and make it available to the new healthcare institution. This process may lead to loss of data or data inconsistency. But this can be prevented with the help of smart contracts; the data can be stored on a blockchain over a distributed ledger.

If the data is stored on a blockchain, it can be made directly available to all other hospitals where a patient is seeking treatment. As every member connected in the chain has a complete copy of the full medical record of an individual, loss of data or corruption of data or a threat of malicious attack cannot result in the loss of data as it can be retrieved from any other block connected in the blockchain. Also, any updating in the record is broadcasted to the entire network, hence resulting in proper maintenance of data integrity of the patients' information.

14.1.4 GAS OPTIMIZATION TECHNIQUES

Smart contract execution consumes some cost which is calculated in gas or gwei. This section presents a collection of optimization techniques used for optimizing smart contract cost. Various optimizing techniques have been discussed [12]. These various techniques are as follows:

1. By changing data types
2. Use contract's byte code for storing data
3. Packing variables as a single unit through the SOLC
4. Packing variables as a single unit with assembly
5. By using function parameters concatenation method
6. By using Merkle proofs to reduce storage load
7. By using Stateless contracts in place of stateful contracts
8. Data stored on InterPlanetary File System (IPFS)

Some other techniques for optimizing smart contracts are also discussed [13–16]. In this study, 24 different methods were discussed, which can be further categorized into five different categories, namely,

1. External transactions,
2. Storage,
3. Saving space,
4. Operations,
5. Miscellaneous.

External Transactions: The methods are related to smart contract creation and sending transactions using external addresses. For example, proxy, data contract, event log, etc.

Storage: These methods are related to storing permanent data on storage. For example, limit storage, packing variables, packing booleans, etc.

Saving Space: These methods are related to saving space in storage as well as in memory. For example, mapping vs. array, Uint vs. Uint256, fixed size, default values, minimize on chain data.

Operations: These methods are related to the gas-consuming operations used in smart contract functions. For example, limit external calls, internal function calls, fewer functions, use libraries, short circuits, write values, limit modifiers, etc.

Miscellaneous: Those methods that are not categorized in any of the above methods are included in this category. For example, freeing storage, optimizer, etc.

14.2 LITERATURE SURVEY

The healthcare industry is losing billions of dollars of amount every year because of incorrect information or data. To reduce the loss or prevent, a system was proposed [17] to handle the activities by storing the data on the blockchain in a secured, transparent, and tamper-proof manner. The data is available for everybody, and records of all the transactions carried out or modifications done within the network are visible to all. This proposed work presented the front-end as well as back-end. The front-end used Redux and React js. and the back-end used Flask and BigchainDB for development.

Another proposed model named Smart-Health or SHealth [18] is a blockchain-based health management system. The government uses it for synchronizing the citizens' health data. It is a fully computerized health management system with high security. SHealth uses smart contracts to manage the patients' requirements such as appointment booking, medical tests booking, and access to the patient details. SHealth architecture has its stack of three layers:

1. The first layer is the government.
2. The second layer is the users.
3. The last layer is the IoT device layer.

The functionality of each layer can be defined as follows: The first layer is the government layers which control access to the blockchain while adding new users to the network. The second layer is the user layer who is the only user already registered by the government layer. The third layer is the IoT terminals layer, which hosts all the medical devices and instruments required for the project. All the registered stakeholders can use the Healthcare system securely for accessing records.

A blockchain-based framework is proposed [19] for an insurance company. All the transactions of an insurance company were being done securely within the stipulated time by using the proposed blockchain-enabled model. The model designated an agent who processes all the clients' requests to the system on their behalf. All the transactions are gathered and saved into the distributed ledger. Before saving onto the blockchain smart contract's validators or supporters can validate or reject these transactions.

An Ethereum-based blockchain framework using robotics was proposed for certificate verification systems [20]. The proposed work presents a smart contract for storing certificates on the blockchain using robotics.

A robotic device was used to scan and store the certificates on the cloud using the Arduino Uno device. Further, the same certificates are stored permanently on the blockchain. It enables the job seekers to upload their certificates on blockchain and enables the employer to verify the stored certificate easily.

Blockchain technology is being widely used in financial applications in the form of cryptocurrencies. The paper by Le Nguyen [21] discusses how this cryptocurrency can be used in making payments and how the process is different from the traditional model of the payment system. Further, a chain of blocks is used to store and share the medical data securely among different stakeholders of the system. These blocks also maintain the transactions related to payments by using smart contracts, storing and sharing the sensitive medical data of the patient and maintaining the transactions of the payment made.

The work by Mohanta et al. [22] is about smart contracts that are self-executed programs used in blockchain. They have tampered proof as cannot be changed once written and executed. They eliminate third-party involvement, making the system automated. They are being used in various blockchain-based applications such as health care, financial, and smart grid to reduce the intermediary involvement as well as to make the system transparent.

A case study on dApp for smart health named DASH [23] has been discussed, a blockchain-based distributed application in the healthcare domain. DASH is a portal that patients can use for accessing self-reports and medical records. The Healthcare system staff can access these reports as well and they are allowed to change the records wherever required. As the technology provides transparency, every stakeholder can see this change of the system.

A study [24] presents how an insurance company can detect frauds regarding premium calculation with the help of this blockchain technology. Further blockchain can also be used for activating/deactivating policies and covers and also help in the decentralization of the system. Blockchain is also helpful in keeping track of all the transactions between the two parties, hence reducing the risk of tampering. A study [25] presented a hierarchical framework for a smart Healthcare system. This system used fuzzy set theory to remove unnecessary attributes. The study used an Imperative Structure Model (ISM) to maintain the hierarchy in the proposed framework.

A smart contract is a core part of the blockchain and is used in the Healthcare system, mainly dependent on record management. The proposed hierarchical framework is envisioned to move the smart Healthcare system towards the use of blockchain while the stakeholders are encouraged to invest in the development of such applications. A study [26] presents a brief overview of different blockchain applications in the healthcare domain. The underlying technology ensures data integrity as well as restoring stakeholder's consensus. Blockchain-based frameworks can be developed to carry out a variety of healthcare applications such as generating and verifying the proof of disease and monitoring the drug supply chains. A conceptual model presented different aspects of various interrelated technologies such as IoT, AI, and blockchain technology to fulfill the needs of the healthcare domain.

SmartInspect [27] is proposed for making debugging process of smart contracts easier. Programming language Solidity is used for writing smart contracts on Ethereum. As smart contracts are once deployed, they are not allowed to change. Debugging is a difficult task in the smart contract as they are embedded in functions with their data. Once smart contracts are deployed and executed, they cannot be re-executed. Once data is encoded, detection of a bug becomes difficult. Hence, here, SmartInspect plays an important role and allows the creators of smart contracts to understand its state and visualize it better without the need to redevelop the code.

A study [28] presents an optimization tool 'GASOL', a cost model for different types of gas consumption for EVM instructions. These EVM instructions are consumed gas that depends upon the instruction's storage space. The tool is a plugin for solidity and displays the gas-instructions pair for a gas-optimized solidity function wherever applicable.

In [29] authors identified seven gas costly patterns and categorized them into three. Apart from that, they proposed a tool 'GASPER' to analyze gas costly patterns. They analyzed 4,240 real smart contracts for three types of patterns and observed that 80% of smart contracts suffered from these types of gas costly patterns.

14.3 METHODOLOGY

The methodology is divided into three parts. These three parts are explained as follows.

14.3.1 Designing of Healthcare System Smart Contract on Ethereum Blockchain

Design can be application-specific. The basic design structure of the Healthcare system smart contract is shown in Table 14.1. Later, it has been improved as per the optimization techniques applied. It has mainly four components:

   i. Name of the smart contract
   ii. Data and state variables
   iii. Functions
   iv. Constructor

The name of the smart contract is Healthcare system. As the name suggests it is related to the Healthcare system . String, bytes32, int256, and address are different data and state variables that have been used in the smart contract. Mainly two functions sendpatientQuery() and getpatientQuery() have been used. The two functions sendpatientQuery() and getpatientQuery() are used to store and retrieve the patient information regarding status, age, and gender.

14.3.2 Writing Healthcare System Smart Contract Using Sublime Text 3 and Remix Ethereum IDE

Smart contracts are written in different languages such as solidity, C++, Java, JavaScript, and Golang. In this work, solidity is used as a language of the smart contract. To develop a smart contract for Ethereum-based project, Sublime Text 3 editor has been used. This written smart contract can be tested in Remix Ethereum IDE. Four different versions of smart contract Healthcare system have been written as shown in Figures 14.3–14.6. In this work, we have tried to improve the smart contract in optimizing the same; hence, four different versions of smart contract code are there (Figure 14.3 is the least optimized smart contract, and Figure 14.6 is the most optimized).

TABLE 14.1
Designing of Healthcare system Smart Contract

Smart Contract Name	Healthcare system
Data and State variables	Address and string types data and state variables
Functions	Two functions: sendpatientQuery() getpatientQuery ()
Constructor	constructor ()

```
pragma solidity ^0.4.22;

contract EnquirySmartContract{

  address public ownerAddress;

  string public patientStatus;

  string public patientAge;

  string public patientGender;

  constructor() public{

    ownerAddress=msg.sender;

  }

  function sendpatientQuery(string _patientStatus,string _patientAge,string _patientGender)public returns(bool)

  {

  patientStatus=_patientStatus;

  patientAge=_patientAge;

  patientGender=_patientGender;

  return true;

  }

  function getpatientQuery () public view returns(address _ownerAddress, string _patientStatus,string _patientAge,string _patientGender)

  {

  _ownerAddress=ownerAddress;

  _patientStatus=patientStatus;

  _patientAge=patientAge;

  _patientGender=patientGender;

  }

}
```

FIGURE 14.3 Smart contract code version 1.

First line in the code pragma Solidity ^0.4.22 presents the version of solidity. This line interprets that one needs to load 0.4.22 or a higher version of solidity to compile and run the smart contract in Remix. Different data and state variables such as string, bytes32, int256, and address are also mentioned in the smart contracts. Change in data variables from the first version of the smart contract to the

```solidity
pragma solidity ^0.4.22;
contract EnquirySmartContract{
    address public ownerAddress;
    string public patientStatus;
    int public patientAge;
    string public patientGender;
    constructor() public{
        ownerAddress=msg.sender;
    }
    function sendpatientQuery(string _patientStatus,int _patientAge,string _patientGender)public returns(bool)
    {
        patientStatus=_patientStatus;
        patientAge=_patientAge;
        patientGender=_patientGender;
        return true;
    }
    function getpatientQuery () public view returns(address _ownerAddress, string _patientStatus,int _patientAge,string _patientGender)
    {
        _ownerAddress=ownerAddress;
        _patientStatus=patientStatus;
        _patientAge=patientAge;
        _patientGender=patientGender;
    }
}
```

FIGURE 14.4 Smart contract code version 2.

```solidity
pragma solidity ^0.4.22;
contract EnquirySmartContract{
    address public ownerAddress;
    bytes32 public patientStatus;
    int public patientAge;
    bytes32 public patientGender;
    constructor() public{
        ownerAddress=msg.sender;
    }
    function sendpatientQuery(bytes32 _patientStatus,int _patientAge,bytes32 _patientGender)public returns(bool)
    {
    patientStatus=_patientStatus;
    patientAge=_patientAge;
    patientGender=_patientGender;
    return true;
    }
    function getpatientQuery () public view returns(address _ownerAddress, bytes32 _patientStatus,int _patientAge,bytes32 _patientGender)
    {
      _ownerAddress=ownerAddress;
      _patientStatus=patientStatus;
      _patientAge=patientAge;
      _patientGender=patientGender;
    }
}
```

FIGURE 14.5 Smart contract code version 3.

```solidity
pragma solidity ^0.4.22;
contract EnquirySmartContract{
  address public ownerAddress;
  struct Details
  {
    bytes32 patientStatus;
    int patientAge;
    bytes32 patientGender;
  }
  Details public details;
  constructor() public{
    ownerAddress=msg.sender;
  }
  function sendpatientQuery(bytes32 _patientStatus,int _patientAge,bytes32 _patientGender)public returns(bool)
  {
    details.patientStatus=_patientStatus;
    details.patientAge=_patientAge;
    details.patientGender=_patientGender;
    return true;
  }
  function getpatientQuery () public view returns(address _ownerAddress, bytes32 _patientStatus,int _patientAge,bytes32 _patientGender)
  {
    _ownerAddress=ownerAddress;
    _patientStatus=details.patientStatus;
    _patientAge=details.patientAge;
    _patientGender=details.patientGender;
  }
}
```

FIGURE 14.6 Smart contract code version 4.

fourth version of the smart contract reduced the gas cost and therefore optimizing the smart contract. Two functions sendpatientQuery() and getpatientQuery() are used to store and retrieve the patient information regarding status, age, and gender. The first version of the smart contract uses only string and address data type. The later versions use bytes32 and int256 data types which will consume less gas cost hence optimize the smart contract. The last version uses package variable structure using the struct keyword. One constructor verifies that the message sender is the owner itself. Data and state variables contained by the Healthcare system smart contract are as follows:

i. ownerAddress – stores owner address of healthcare smart contract
ii. patientStatus – stores the patient's status
iii. patientAge – stores the patient's age
iv. patientGender – stores the patient's gender

Functions contained by the Healthcare system smart contract are as follows:

Constructor () public {ownerAddress=msg.sender;} – As the smart contract is executed, it verifies whether the message sender is smart contract owner or not.

function sendpatientQuery (string _patientStatus, string patientAge, string_ patientGender) – A new patient's details are recorded as the function is called.

function getpatientQuery public view returns (address _ownerAddress, bytes32 _patientStatus, int _patientAge, bytes32 _patientGender) – A patient's details are fetched as the function is called.

After writing Healthcaresystem smart contract, this needs to be compiled as well as tested on Remix Ethereum IDE. Compilation details of the smart contract are mentioned in the artifacts generated at the compilation time and stored in a JSON file. The compiled contract details are presented in Figure 14.7.

Different elements of the artifacts are name, Application Binary Interface (ABI), metadata, web3 deploy, Byte code, metadata hash, swarm location, etc. Some of them are described as follows:

Contract Application Binary Interface (ABI): It contains all functions of the smart contract with their parameters and returns values.
Contract Bytecode: This tells about the smart contract data that has already been compiled.
Web3 Deploy Script: JavaScript Object Notation (JSON) file contains the script that is used for invoking functions and for deploying the smart contract.
Gas Estimates: It provides the details of gas consumption of running a smart contract.
Assembly Code: Every instruction is written in the smart contract has its equivalent operational codes.

After finishing the compilation process, the environment is set as Javascript Virtual machine (VM), and one account is chosen from the available 10 default accounts. The smart contract is then deployed using Remix Ethereum IDE (Figure 14.8).

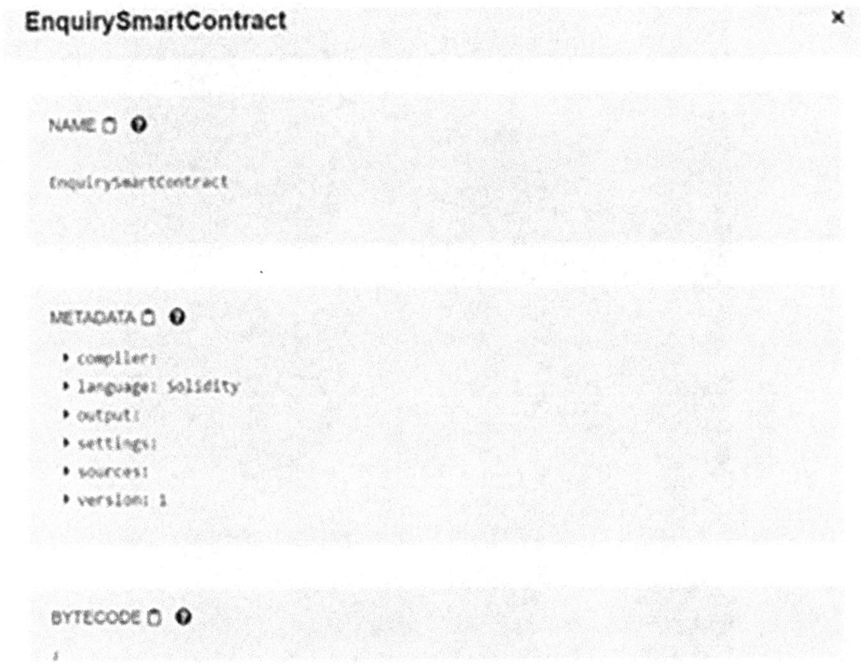

```
EnquirySmartContract                                    ×

    NAME

    EnquirySmartContract


    METADATA
      • compiler:
      • language: Solidity
      • output:
      • settings:
      • sources:
      • version: 1


    BYTECODE
```

FIGURE 14.7 Artifacts generated for optimized smart contract.

14.3.3 OPTIMIZATION TECHNIQUES APPLIED ON SMART CONTRACT TO REDUCE THE GAS COST

The basic structure of the smart contract is given in Figure 14.3 in which only string data type is used for all variables. Further, we tried to optimize the gas cost by using different data types such as bytes32 for patientStatus and patientGender and int256 for patientAge. Finally, we used a packaged variable struct to combine all three variables into a single unit. The reduced amount of gas cost will be discussed in the Results section in Table 14.2.

14.4 RESULTS

Various optimization techniques were used to reduce the gas consumption of the smart contract. Analysis of gas consumption is given in Table 14.2.

Four cases for optimization have been shown in Table 14.2. For all cases, the gas limit is the same which is 3000000 (Gwei).

1. All the variables are of string type. No optimization is applied in this case. As mentioned in row one, the transaction cost is 22,800 gas, and the execution cost is 1,528 gas. As optimization is not applied, hence no reduction in both costs.

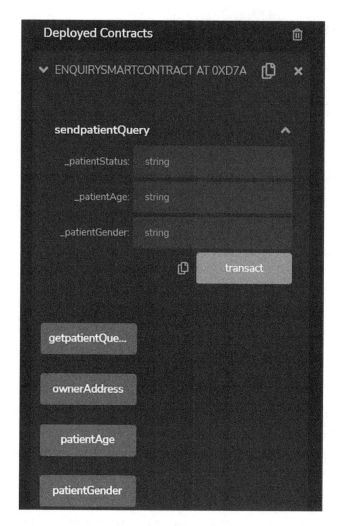

FIGURE 14.8 Snapshot of deployed smart contract.

2. In this case, which is shown in the second row of the table, int256 data type has been used in place of string data type for patientAge variable. Transaction cost is calculated as 22,756 gas and execution cost is 1,484 gas. Only one change in data type reduced transaction cost by 0.19% and execution cost by 2.87%.
3. In this case, which is shown in the third row of the table, int256 data type has been used in place of string data type for patientAge variable; both variables patientStatus and patientGender changed their data type from string to bytes32. Transaction cost is calculated as 21,750 gas and execution cost is 478 gas. This change in data type reduced transaction cost by 4.60% and execution cost by 68.71%.

TABLE 14.2
Observed Cases for Gas Cost by Applying Different Optimization Techniques

S.No.	Data Type Used	Optimization	Gas Limit (Gwei)	Transaction Cost	Execution Cost	% Reduction in Transaction Cost	% Reduction in Execution Cost
1.	String data type for patientStatus, patientAge, and patientGender variables	Not Applied	3,000,000	22,800 gas	1,528 gas	NA	NA
2.	String data type for patientStatus and patientGender variables, int256 for patientAge variable	Applied	3,000,000	22,756 gas	1,484 gas	0.0019	2.87%
3	bytes32 data type for patientStatus and patientGender variables, int256 for patientAge variable	Applied	3,000,000	21,750 gas	478 gas	4.60%	68.71%
4	Packed data type struct combined three variables: bytes32 patientStatus, bytes32 patientGender, and int256 patientAge into a single unit	Applied	3,000,000	21,780 gas	508 gas	4.47%	66.75%

4. In this case, which is shown in the fourth row of the table, all the variables patientAge, patientStatus, and patientGender have been combined into a single unit using struct data type. Transaction cost is calculated as 21,780 gas and execution cost is 508 gas. This change in data type reduced transaction cost by 4.47% and execution cost by 66.75%.

Overall, we can say transaction cost as well as execution cost reduced in each case.

14.5 CONCLUSION

This work presents a blockchain-based implementation of the healthcare system as a smart contract. A smart contract has been created which can be used for speeding up the process as well as eliminating human involvement by effectively digitizing the process. The work comprises three steps from designing the smart contract to optimize the smart contract. This will help the user to manage the healthcare system. Smart contracts enable the user to record patient details transparently and eliminating the possibility of data manipulation in the future. The work clearly showed that the smart contract has been created on Ethereum, used solidity as a language, and executed on Remix Ethereum IDE which is an online compiler.

The basic structure of the smart contract is described above, in which only string data type is used for all variables. Further, we tried to optimize the gas cost by using different data types such as bytes32 for patientStatus and patientGender and int256 for patientAge. Finally, we used a packaged variable struct to combine all three variables into a single unit, hence reducing the amount of gas cost. All the variables are of string type. When no optimization is applied, the transaction cost is 22,800 gas, and the execution cost is 1,528 gas. When int256 data type is used in place of string data type for patientAge variable, the transaction cost is calculated as 22,756 gas and execution cost as 1,484 gas. Only one change in data type reduced transaction cost by 0.19% and execution cost by 2.87%. When both variables patientStatus and patientGender changed their data type from string to bytes32, the transaction cost is calculated as 21,750 gas and execution cost as 478 gas. This change in data type reduced transaction cost by 4.60% and execution cost by 68.71%. When all the variables patientAge, patientStatus, and patientGender have been combined into a single unit using struct data type, the transaction cost is calculated as 21,780 gas and execution cost as 508 gas. These changes in data types reduced transaction costs by 4.47% and execution cost by 66.75%. Overall, we can say transaction cost as well as execution cast reduced in each case.

This work is specific for managing the healthcare system, and it can further be modified according to the needs and requirements of the application. As the patient details are uploaded on the blockchain which is a public blockchain, the medical details of an individual that are confidential might create some objection when it's displayed on a public blockchain. This approach can further be implemented on a private blockchain where only the stakeholders can access the data. This will help in maintaining confidentiality and increasing security as well as maintaining privacy.

REFERENCES

1. Nofer, M., Gomber, P., Hinz, O., & Schiereck, D. (2017). Blockchain. *Business & Information Systems Engineering*, 59(3), 183–187. Doi: 10.1007/s12599-017-0467-3.
2. Wang, S., Ouyang, L., Yuan, Y., Ni, X., Han, X., & Wang, F.-Y. (2019). Blockchain-enabled smart contracts: architecture, applications, and future trends. *IEEE Transactions on Systems, Man, and Cybernetics: Systems*, 1–12. Doi: 10.1109/tsmc.2019.2895123.
3. Nofer, M., Gomber, P., Hinz, O., & Schiereck, D. (2017). Blockchain. *Business & Information Systems Engineering*, 59(3), 183–187. Doi: 10.1007/s12599-017-0467-3.
4. Kumar, T., Ramani, V., Ahmad, I., Braeken, A., Harjula, E., & Ylianttila, M. (2018). Blockchain utilization in healthcare: key requirements and challenges. *2018 IEEE 20th International Conference on e-Health Networking, Applications and Services (Healthcom)*. Doi: 10.1109/healthcom.2018.8531136.
5. Kassab, M. H., DeFranco, J., Malas, T., Laplante, P., Destefanis, G., & Neto, V.V.G. (2019). Exploring research in blockchain for healthcare and a roadmap for the future. *IEEE Transactions on Emerging Topics in Computing*, 1–1. Doi: 10.1109/tetc.2019.2936881.
6. Wohrer, M., & Zdun, U. (2018). Smart contracts: security patterns in the ethereum ecosystem and solidity. *2018 International Workshop on Blockchain Oriented Software Engineering (IWBOSE)*. Doi: 10.1109/iwbose.2018.8327565.
7. Alharby M., and Van Moorse, A. (2017) Blockchain-based Smart Contracts: A Systematic Mapping Study of Academic Research (2018), *2018 International Conference on Cloud Computing, Big Data and Blockchain (ICCBB)*, IEEE, Fuzhou, pp. 1–6. Doi: 10.1109/ICCBB.2018.8756390.
8. Kõlvart, M., Poola, M., & Rull, A. (2016). Smart contracts. *The Future of Law and eTechnologies*, 133–147. Doi: 10.1007/978-3-319-26896-5_7.
9. Zheng, Z., Xie, S., Dai, H. N., Chen, X., & Wang, H. (2018). Blockchain challenges and opportunities: a survey. *International Journal of Web and Grid Services*, 14(4), 352. Doi: 10.1504/ijwgs.2018.095647.
10. Christidis, K., & Devetsikiotis, M. (2016). Blockchains and smart contracts for the internet of things. *IEEE Access*, 4, 2292–2303. Doi: 10.1109/access.2016.2566339.
11. Idelberger, F., Governatori, G., Riveret, R., & Sartor, G. (2016) Evaluation of logic-based smart contracts for blockchain systems, pp. 167–183.
12. Lucas A. 8 ways of reducing the gas consumption of your smart contracts. Available at, https://medium.com/coinmonks/8-ways-of-reducing-the-gas-consumption-of-your-smart-contracts-9a506b339c0a, last access on May 2021.
13. Eattheblocks. How to optimize gas cost in a Solidity smart contract? 6 tips. Available at: https://eattheblocks.com/how-to-optimize-gas-cost-in-a-solidity-smart-contract-6-tips/, last access on December 2019.
14. Gupta, M. Mudit Gupta's Blog. Available at: https://mudit.blog/solidity-gas-optimization-tips/, last access on December 2019.
15. Gupta, M. Solidity tips and tricks to save gas and reduce bytecode size. Available at: https://blog.polymath.network/solidity-tips-and-tricks-to-save-gas-and-reduce-bytecode-size-c44580b218e6, last access on December 2019.
16. Shahda, W. Gas optimization in solidity part I: variables. Available at: https://medium.com/coinmonks/gas-optimization-in-solidity-part-i-variables-9d5775e43dde, last access on December 2019.
17. Saldamli, G., Reddy, V., Bojja, K. S., Gururaja, M. K., Doddaveerappa, Y., & Tawalbeh, L. (2020). Health care insurance fraud detection using blockchain. *2020 Seventh International Conference on Software Defined Systems (SDS)*. Doi: 10.1109/sds49854.2020.9143900.

18. Zghaibeh, M., Farooq, U., Hassan, N. U., & Baig, I. (2020). SHealth: a blockchain-based health system with smart contracts capabilities. *IEEE Access*, 1–1. Doi: 10.1109/access.2020.2986789.
19. Raikwar, M., Mazumdar, S., Ruj, S., Sen Gupta, S., Chattopadhyay, A., & Lam, K.-Y. (2018). A blockchain framework for insurance processes. *2018 9th IFIP International Conference on New Technologies, Mobility and Security (NTMS)*. Doi: 10.1109/ntms.2018.8328731.
20. Malsa, N. Vyas, V., Gautam, J., & Shaw, R.N. Framework and smart contract for blockchain enabled certificate verification system using robotics. *Machine Learning for Robotics Applications*, 125–138.
21. Le Nguyen, T. (2018). Blockchain in healthcare: a new technology benefit for both patients and doctors. *2018 Portland International Conference on Management of Engineering and Technology (PICMET)*. Doi: 10.23919/picmet.2018.8481969.
22. Mohanta, B. K., Panda, S. S., & Jena, D. (2018). An overview of smart contract and use cases in blockchain technology. *2018 9th International Conference on Computing, Communication and Networking Technologies (ICCCNT)*. Doi: 10.1109/icccnt.2018.8494045.
23. Zhang, P., Schmidt, D. C., White, J., & Lenz, G. (2018). Blockchain technology use cases in healthcare. *Advances in Computers*. Doi: 10.1016/bs.adcom.2018.03.006.
24. Kumar, T., Ramani, V., Ahmad, I., Braeken, A., Harjula, E., & Ylianttila, M. (2018). Blockchain utilization in healthcare: key requirements and challenges. *2018 IEEE 20th International Conference on e-Health Networking, Applications and Services (Healthcom)*. Doi: 10.1109/healthcom.2018.8531136.
25. Du, X. Chen, B. Ma, M. & Zhang, Y. (2021). Research on the application of blockchain in smart healthcare: constructing a hierarchical framework. *Journal of Healthcare Engineering*, 2021, Article ID 6698122, 13 p. Doi: 10.1155/2021/6698122.
26. Shukla, R.G., Agarwal, A., & Shukla, S. (2020). *Blockchain-Powered Smart Healthcare System, Handbook of Research on Blockchain Technology*. Academic Press. ISBN 9780128198162. Doi: 10.1016/B978-0-12-819816-2.00010-1.
27. 1 GEO. L. TECH. REV. 273 (2017). https://perma.cc/TY7W-Q8CX.
28. Albert, E., Correas, J., Gordillo, P., Román-Díez, G., & Rubio, A. (2020, April). GASOL: gas analysis and optimization for ethereum smart contracts. *International Conference on Tools and Algorithms for the Construction and Analysis of Systems*, pp. 118–125, Springer, Cham.
29. Chen, T., Li, X., Luo, X., & Zhang, X. (2017, February). Under-optimized smart contracts devour your money. *2017 IEEE 24th International Conference on Software Analysis, Evolution and Reengineering (SANER)*, pp. 442–44, IEEE.

15 Interference Mitigation Using Cognitive Femtocell from 5G Perspective

Gitimayee Sahu and Sanjay S. Pawar
SNDT Women's University

CONTENTS

15.1 Introduction 251
15.2 Motivation 253
15.3 Objective 254
15.4 Literature Review 254
15.5 System Model 256
    15.5.1 Functions of CFC 258
    15.5.2 Biasing and SINR 258
    15.5.3 Minimizing Interference 259
    15.5.4 The Resource Allocation Process 260
15.6 Results and Discussion 263
15.7 Conclusion 265
References 266

15.1 INTRODUCTION

The concept of cognitive radio was popularized by J. Mitola in 1999 [1]. The cognitive concept developed on software-defined radio (SDR) is termed as an inventive cellular communication system [1]. That is familiar with its surroundings, and it utilizes the concept of learning from the environment. It adapts the statistical deviations in the input stimulant with two fundamental concepts; highly impeccable communication everywhere and adequate utilization of radio resources. The cognitive radio can able to (i) Observe: sense the environment, (ii) Plan: design various scenarios, (iii) Decide: choose the best possible action, (iv) Learn: derive its behavior, and (v) Act: provide the communication.

Cognitive femtocell (CFC) includes the cognitive principle in a two-tier heterogeneous network (HetNet) and allows cost-effective deployment of the femtocell. It provides network awareness, spectrum sharing, and active radio resource management. Network awareness enables cognizance both at radio access networks (RAN) and user devices. It facilitates the realization of many aspects of the network, such as topology, traffic behavior, characteristics and constraints, load upon the network, location of the base station (BS) & users (UEs), and available radio resources. The sensing mechanism uses an energy detection technique to detect the available spectrum chunks. The frequency channels allocated to cell edge UEs, i.e. far from the BS, can be reused by the femtocell. The resource blocks (RBs) can be used orthogonally in order to avoid cross-tier interference between the macro base station (MBS) and femtocell. The interference may reduce signal to interference noise ratio (SINR) of the user, throughput, and spectral efficiency of the overall network. The CFC senses the available spectrum, dynamically assigns the vacant channels to the users, and maximizes the network utility. There are two objectives: honor and cost. The honor is the reward in terms of gain accomplished by certain UEs while selecting a specific channel. The cost in terms of inter and intra-channel interference in the HetNet. While spectrum sharing between macro UEs (MUEs) may experience low performance due to path loss and interference. Femtocell UEs (FUEs) receive high-quality signal since the distance between the UEs and the access points are minimum. The intra-tier interference between the femtocell and inter-tier interference between macrocell and femtocell should be less for high performance of the network.

The 5G technology offers 10 Gbps peak data rate, 1 ms latency, 100 times more connected devices than 4G, high availability, longer battery life for low power devices, and 90% reduction in energy consumption, hence highly energy-efficient [2,3]. To support these features small cells (i.e., picocell, microcell, and femtocell, i.e. home eNodeB (HeNB)) have to be deployed under the macrocell coverage. Hence the network gets heterogeneous and ultra-dense in architecture. 5G network has cognitive and co-operative capability for proper utilization of radio resources to increase the spectrum efficiency of the network. The prime technologies include cognitive radio (CR) and cognitive cycle (CC). CR enables the nodes to find the underutilized licensed channels. CC has been encapsulated in CR nodes to learn new knowledge and adopt the network dynamics. The CR and CC have brought advantages to the cognitive RAN network considering the issue of spectrum scarcity. It enhances the interoperation among the network entities and provides perception and sovereign capability of 5G core operation. The step-by-step procedure of the CC is represented in Figure 15.1. It consists of network awareness, selection, and access procedure. The network awareness includes sensing, analysis, and learning. It senses the environment, gathers information, analyzes and learns from the environment, and then decides to configure and adopt the parameters of radio network. The CR and CC cycles can be used in 5G technology for enhancing energy efficiency, spectral efficiency, cost efficiency, and increasing quality of service (QoS) [4–6].

The vision of 5G can be realized through cloud RAN (C-RAN) [7–9]. C-RAN is an emerging concept and one of the most imperative architectures to satisfy the demands of 5G. In essence, it accommodates next-generation network such as statistical multiplexing, resource slicing, energy efficiency, and high capacity. C-RAN infrastructure includes software-defined network (SDN) and network function virtualization (NFV). SDN separates the control plane and data plane and offers load balancing between them. The concept of NFV offers high flexibility by allowing network resource sharing in a dynamic way [10].

Interference Mitigation Using Cognitive Femtocell

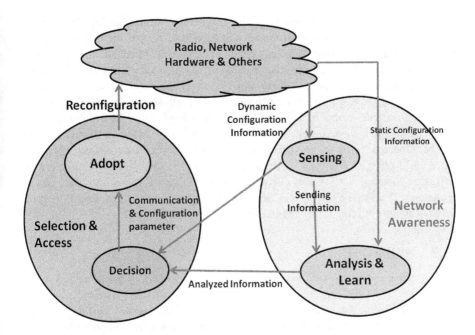

FIGURE 15.1 Various stages of cognitive cycle.

To implement C-RAN, switches act as base band units (BBUs) in the pool. The pool coordinator is the controller which controls the BBUs. Resource allocation has to be done between BBUs and remote radio units (RRUs) with high energy efficiency. The host manager manages the BBU pool monitoring load on each BBU. It sends the BBU to sleep mode which is not in use for enhancing energy efficiency. It minimizes delay, less complex, provides good QoS, and maintains proportional fairness among users. C-RAN introduces a new layer in mobile network known as front haul, which is adopted from the recent research works in fiber wireless paradigm. The front haul defines the link between the BBU and the RRU. Different technologies such as millimeter-wave and optical fiber can be used for this purpose (Figure 15.1).

The chapter is arranged in the following manner. Section 15.1 explains the introduction, Section 15.2 explains Motivation, Section 15.3 Objective, and Section 15.4 Literature Review in concise. Section 15.5 explains the System Model, Section 15.6 Results and Discussion, and Section 15.7 explains the Conclusion.

15.2 MOTIVATION

5G network has heterogeneous architecture that accommodates different types of the BS and users of various service classes. This heterogeneity makes the network ultra-dense; due to this, intra- and inter-tier interference raises which degrades the SINR, capacity, and performance of the network as well. As the number of mobile subscribers increases exponentially year-wise, i.e. nearly 5.7 billion (71% of total population) by 2023 [11], to get guaranteed QoS, efficient resource allocation is essential. The spectrum resources are scarce, and a large number of users are there. Thus the cognitive concept of SDR should be adopted to sense and find the whitespace in TV spectrum bands. The database of the cash memory at each BS needs to be updated on real time, so that the idle RBs or frequency channels can be identified and allocated to the demanding users on space time-frequency basis. 5G technology promises for higher data rate, ultra-low latency, and massive machine type communication. So to achieve these use cases the distance between the user and the BS should be less. Femtocell or HeNB with cognitive concept, i.e. CFC, is a suitable implication for it commemorates high data and high capacity. The CFC can sense neighbor BSs for spectrum holes that can be allocated to the FUEs. The nearby femtocells form cluster known as coordinated multi-point (CoMP), so that the users traveling in same cluster will not perform handover. When the MBS gets congested in peak hours, some of the users can be biased to offload to the nearby CFC for load sharing, thus enhancing efficiency of the network by minimizing intra- and inter-tier interference.

15.3 OBJECTIVE

The objective of the research work is to combine the concept of fractional frequency reuse (FFR) and CFC for minimizing interference and adequate resource allocation. In the FFR technique, the center area of the cell will be served by the MBS and the cell edge area served with CFC with different frequencies compared to the cell center. The signal strength of MBS reduces towards the cell boundary. The SINR decreases and the user may enter the probability of outage. If the cell edge users are served with femtocell clusters, the user will get the requisite QoS since the distance between BS and user reduces. The CFCs sense and explore the spectrum which is not in use by the primary user (PU). Because of this, the intra- and inter-tier interference reduces and increases the performance of the network.

15.4 LITERATURE REVIEW

Resource allocation becomes more challenging in C-RAN. It becomes a multi-objective optimization problem. The various objectives are (i) cost, (ii) delay, (iii) power consumption, and (iv) complexity.

In cost-efficient resource allocation, the entire resource is divided into fixed and shared manner. When the network is congested, the shared resources are mutually distributed among the low priority users. When the network is not loaded, the fixed or dedicated resources can be assigned to high priority users. In this manner, the utility can be maximized with proper allocation of resources even in highly congested scenarios. It satisfies the user demand. The users can be prioritized depending on how much price they are paying. The central controller allocates the resources depending on the user priority, amount of waiting delay, and type of services. The different services include (i) Real-time calls (high priority); (ii) Interactive services, e.g. streaming, chatting, email, and online gaming (medium priority); and (iii) Background services, e.g. downloading (low priority). Resource allocation using power minimization strategy selects the nearest RRU with low power consumption and multiple RBs. In this manner inter-channel interference reduces and the user achieves high throughput with guaranteed QoS.

Xiaoyu Wang et al. in [12] propose cognitive empowered femtocell for dynamic spectrum management of femtocell network. It mitigates both cross-tier and intra-tier interference. The cognitive-enabled femto base stations (FBS) and FUEs use the spatio-temporarily available radio resources. Li Huang et al. in [13] explain co-channel deployment of closed subscriber group femtocell causing co-channel interference, leading to coverage hole. There three different CR-enabled interference mitigation techniques are discussed, i.e. opportunistic interference avoidance, interference cancellation, and interference alignment. A joint interference avoidance scheme with Gale-Shapley spectrum sharing (GSOIA) based on an interweaving paradigm to mitigate intra- and inter-tier interference in macro/femto HetNet [14]. CFC opportunistically communicates over the available spectrum with the least possible interference to the macrocell. Different femtocells are assigned with orthogonal spectrum resources to minimize intra-tier interference [15]. Saba et al. in [16] proposed CFC for solving the problem of spectrum scarcity for indoor applications. A broadband router is used for controlling the delivery of data packets between the macro cell and CFC using priority queuing strategy. The solution combines physical, MAC, and application layers to provide the demanded QoS in addition to the adaptive changes in wireless network. Gürkan Gür et al. in [17] explain CFC for dynamic spectrum access.

Guobin Zhang et al. in [18] propose power management using non-convex optimization in CFC for reducing interference. A closed-form expression is derived for optimal power configuration to enhance the capacity and decrease the energy consumption concurrently. Ole Grøndalen et al. in [19] analyze the business aspect of mobile traffic offloading to CFC. Using sensors for searching the unoccupied spectrum, the CFC can find and utilize frequencies other than the mobile network. The business case analyzes the potential for cost savings while offloading the mobile traffic with the femtocell. Instead of using sensors separately, in this work sensors were inbuilt in the CFC itself. It enhances the spectral efficiency and offloading gain of the network. Joydev Ghosh et al. in [20] explain optimization of throughput and network coverage with implications of CFC and beam steering approach. Ayesha Salman et al. in [21] explain spectrum sensing based on near field localization in femtocell network using evolutionary programming such as genetic algorithms. Hesham M. et al. in [22] explain scheduling of power allocation in CFC for hybrid access.

Anggun F. et al. in [23] explain the comparative study of distributed and centralized power control in CFC networks. Centralized power control (CPC) can achieve SINR more than the target SINR, but the power consumption of CPC is much larger than the distributed power control. Hence distributed power control can save more power than centralized control. The dynamic spectrum allocation method can be used for CFC in the downlink direction using orthogonal frequency division multiple access (OFDMA)-based hybrid access [15,24]. The MBS allocates the portion of the sub-channel to the Femtocell Access Point (FAP) to serve the offloaded FUEs. It ensures guaranteed QoS and maximization of throughput of the offloaded UEs. The resource allocation problem is formulated to maximize sum utility and optimization using the dual decomposition method.

Imad et al. in [25] present a heterogeneous cloud radio access network (H-CRAN) and integrate the advantages of HetNet and C-RAN by radio resource management, interference minimization, and scalability. H-CRAN provides co-operative interference mitigation and increases time-sharing between remote radio head (RRH) users. It proposes an absolute blank subframe (ABS) method to improve SINR and throughput of the small cell and MUEs. Jun Wu et al. in [26] explain the logical structure of C-RAN combines the physical plane, control plane, and service plane. The C-RAN architecture uses new communication and computer techniques for service-oriented resource scheduling and management.

Parallel pre-coding and coordinated user scheduling algorithms were used for better performance. Tara et al. in [27] explain the challenges of C-RAN in wireless network virtualization while sharing the resources among BBUs. It performs load balancing in economic network architecture. This minimizes interference and improves the system throughput.

15.5 SYSTEM MODEL

The system model consists of two-tier HetNet comprising of a set of N cognitive femto base stations (CFBS) and a single MBS located at the center, i.e. at (0.5,0.5) of 1 km². area. The macro users (MUEs) are distributed uniformly throughout the area as shown in Figure 15.2. The CFC or, cognitive home eNodeBs (CHeNB) also located uniformly in the area. Each femtocell is exclusively attached with two femto users (FUEs). The MUEs were offloaded from congested MBS to CFC in order to increase the QoS of the users. The whole system is simulated using MATLAB14b. The Poisson point process (PPP) and voronoi tessellation were used with MATLAB14b during the simulation process. The MeNB has a radius of 500 m., and HeNBs were distributed nearly 200 m. at the edge of the MeNB. The downlink transmitter power of MeNB and HeNB were 46dBm and 20dBm, respectively. The pathloss propagation model of MeNB to MUE is $15.3 + 37.6\log_{10} R$ (dB), MeNB to FUE is $15.3 + 37.6\log_{10} R + PL_{hw}$ (dB), and HeNB to FUE is $38.46 + 37.6\log_{10} R + 0.7R$ (dB). Where 'R' is the distance between the eNodeB and the associated user to it and PL_{hw} (=10 dB) is the wall penetration loss. There is 'M' number of sub-channels employed by the MBS. Each CFBS assigned with M_n number of sub-channels doesn't interfere with the other CFBS. Each cognitive femtocell user equipment (CFUE) is assigned with a single sub-channel for data transmission. The central controller is connected to the core network which is used to communicate data between different nodes in the heterogeneous cognitive network.

Macro cell situated at (0.5, 0.5), blue dots represent the femto cells, the green dots show the FUEs, and red dots show the MUEs.

There are three different access mechanisms: (i) open access, (ii) closed access, and (iii) hybrid access. In open access, all the offloaded users can access the femtocell. But in closed access, offloaded users were restricted to use the femto cell, and in hybrid access, the limited number of offloaded MUEs can access the femtocell. In closed access, only a limited number of users are allowed to connect with guaranteed QoS. In open access, all subscribers are allowed to access the femto cell. In hybrid access, all UEs can access but certain UEs are considered as high priority.

Interference Mitigation Using Cognitive Femtocell 257

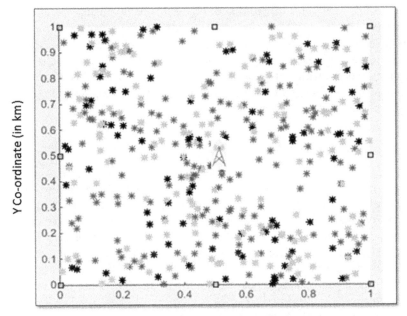

FIGURE 15.2 Two-tier HetNet (macro cell and femto cell).

The three different cognitive scenarios in the network are (i) underlay, (ii) overlay, and (iii) interweave. In underlay, both PU and cognitive user (CU) share the same radio resource by keeping their power level low in order to minimize the interference below threshold. This is mainly used for short range communication. In overlay scenario, PU or the licensed user has the priority of using the frequency channel. Till the completion of data transfer of the PU, the CU has to wait. As soon as the PU completes its transmission, the CU can take control over the channel. But it results in poor performance due to the interference of the PU. Hence this model was discarded by FCC in 2007 [15]. The two cognitive BSs share information in order to obtain prior knowledge about synchronous transmission. In interweave scenario, CU transmits only at the spectrum holes, and as soon as PU comes, the CU has to vacate the channel to reduce the interference.

The different spectrum sensing techniques are (i) energy detection, (ii) cyclostationary detection, (iii) pilot-based coherent detection, and (iv) co-variance detection. Energy detection technique is widely used because of easier implementation and low computational complexity. The radio resources can be used orthogonally to decrease cross-tier interference. But it reduces the spectral efficiency and the co-tier interference which may not be mitigated at all. The cognitive FAP based on the sensing mechanism will dynamically assign available channels to the contesting users.

Under dynamic spectrum sensing mechanism, (i) co-operative sensing and (ii) standalone sensing are two suitable methods extensively used. In co-operative spectrum sensing, the central controller, i.e. base station, cluster head, and data sink controls, schedules sensing of different wireless nodes. It displays the results by integrating the sub-bands which were free to access. With standalone sensing, each secondary node senses the sub-bands in order to know the accessibility.

With coordinated sensing, when there is a need for large radio resources, standalone sensing can be performed with co-ordination of CFC. The sensing co-ordination process with CFC ensures better spectrum usage and adequate interference management. The self-organizing feature of CFC and its compatibility with the existing protocol makes it highly efficient.

15.5.1 Functions of CFC

The CFC and its users inspect the availability of macro cell's licensed bands. It senses the available RBs of Ultra High Frequency (UHF)/Very High Frequency (VHF) TV in regular interval of time. As per the real-time sensing results of available channel information, the CFC schedules the FUEs according to priority. The scheduling is done as per the availability of the sub-bands after sensing in a stochastic manner. The coordinate sensing procedure is highly efficient since each FUE has to scan small sub-bands in a particular sequence. Hence it reduces the delay due to sensing and power consumption. The range and the priority of the channels to be scanned have to be defined.

The end-user module determines the sensing parameters, such as the number of iterations and the sequence of channels for sensing selected sub-bands. As FUEs access the sub-bands, the FUE acknowledges (ACK) the CFC regarding the usage. The ACK message helps the cluster coordinator to perform channel measurement on respective spectrum chunks to analyze the presence of PU, i.e. MUE signal. If any MUE is identified, the cluster coordinator will direct the FUE to vacate the frequency band immediately.

Consider 5G two-tier HetNet where femto cell or HeNB is deployed inside the home and FUEs are associated with it. The nearby femtocells were deployed in clusters. The congestion of the macrocell can be reduced by mobile traffic offloading mechanism. Where the cell edge MUEs are biased to offload to the femtocell, thus it increases energy efficiency of the macro cell and enhances the spectrum utilization of femtocell. This results in increase in spectrum efficiency of femtocell and guaranteed QoS of the user.

The two tiers operate with similar carrier frequency, i.e. (f_c = 2GHz). The MUEs which were positioned in the femto coverage area will receive signals from both macrocell and femtocell. Hence intra-tier interference in the downlink direction increases, which reduces SINR and results in performance deterioration.

15.5.2 Biasing and SINR

For offloading a positive bias (α) is to be added to the received signal strength (RSS) during the cell selection process. The ith user prefers the nth femtocell when α_k biasing is added to the measured RSS from the serving BS.

$$\text{Cell ID}_i = \arg\max_{k}(\text{RSS}_k + \alpha_k) \quad (15.1)$$

Interference Mitigation Using Cognitive Femtocell

where α_k is zero for the macrocell and a positive value for the femto cell. This biases users to choose femtocell as their serving BS gradually. The association probability of ith user connected to the femtocell is

$$p_A = P\{P_f \alpha h_f d_f^{-\eta_f} > P_m h_m d_m^{-\eta_M}\} \tag{15.2}$$

where P_f is the femtocell transmit power, α is the biasing factor, h_f is the channel gain, d_f is the distance between the user and femtocell, η_f is the pathloss factor, P_m is the macrocell transmit power, h_m is the channel gain, d_m is the distance between the user and the macrocell, and η_m is the path loss exponent. The SINR of FUE is given by

$$\Gamma_{f,n} = \frac{P_f h_{f,n}}{\sum_{f'=1, f' \neq f}^{N} P_{f'} h_{f', \{f,n\}} + P_m h_{m\{f,n\}} + \sigma_{f,n}^2} \tag{15.3}$$

where $\sum_{f'=1, f' \neq f}^{N} P_{f'} h_{f', \{f,n\}}$ is the interference induced by the neighboring cells and $P_m h_{m\{f,n\}}$ is the interference to the FUEs by the macro cell. The total interference is given by

$$I = \sum_{f'=1, f' \neq f}^{N} P_{f'} h_{f', \{f,n\}} + P_m h_{m\{f,n\}} \tag{15.4}$$

If the femtocell has cognitive capabilities, i.e. cognitive femtocell, then it can sense the environment; hence interference can be reduced.

15.5.3 Minimizing Interference

The main purpose of femtocell is to provide better indoor coverage and support high bandwidth consuming users with guaranteed QoS. The macrocell signal strength reduces significantly inside the home due to wall penetration loss. Hence to reduce coverage hole and for better services, the HeNB is used. The interference is minimized by adjusting the power levels. As the transmitting power of the femto cell increases, it creates interference to the adjacent femtocell. To meet such types of challenges, power control algorithm [23] is proposed in conjunction with the CFC. The power control algorithm has to be performed with the following steps:

Step I: During the allocation of subcarrier, the cognitive FAP has to sense the channels (using energy detection or cyclo-stationary method) available in macrocell and assigns the channel with minimum interference level.

Step II: The measured SINR indicates the amount of interference in a specified subcarrier.

Step III: There should be a lower threshold of SINR Γ_L and a higher threshold of SINR Γ_H.

If the measured SINR is within the range: $[\Gamma_L < \Gamma_{measured} < \Gamma_H]$, then the subcarrier is allocated to the user.

Step IV: If $\Gamma_{measured} < \Gamma_L$, then adaptive power control of the BS is done to minimize the interference.

Step V: After allocation of the subcarrier, repeat step II to measure the SINR frequently. If variation of SINR occurs, power control of BS is required.

15.5.4 THE RESOURCE ALLOCATION PROCESS

Consider two-tier HetNet consists of macrocell and femtocell. The macrocell having MUE density λ_{MUE} and the femto cell having FUE density λ_{FUE} are shown in Figure 15.3. Assume downlink transmission of orthogonal frequency division multiple access (OFDMA) technique, where the entire bandwidth B is divided into N_f frequency channels (f_1, f_2, f_3) near the cell edge. Each channel is partitioned into a number of RBs. The RB is a space time-frequency unit. The MBS utilizes the

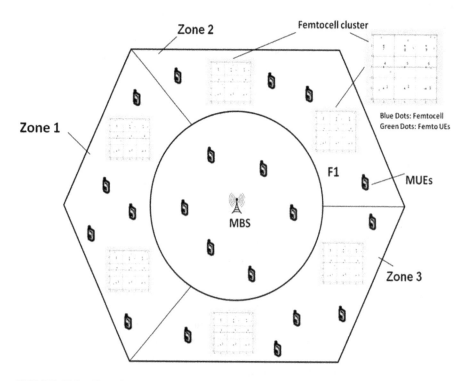

FIGURE 15.3 Two-tier HetNet using FFR technique and clustered femtocell at the cell edge area.

Interference Mitigation Using Cognitive Femtocell

frequency channel in a dedicated and shared manner as well. The hexagonal cell is divided into center and edge regions. At the cell center let the frequency channel be f_c which is different from the cell edge frequencies (f_1, f_2, f_3). The center or the inner region of the adjacent cells can have the same frequency (f_c) which varies from the cell edge frequencies. This technique is known as FFR [28]. The femtocell is represented as $(3 \times 3 = 9)$ cluster at the edge of the macro cell, uniformly distributed at the cell edge areas.

The SINR of FUEs is given by

$$\Gamma_{f,n} = \frac{P_f h_{f,n}}{\sum_{f'=1, f' \neq f}^{N} P_{f'} h_{f',\{f,n\}} + P_m h_{m\{f,n\}} + \sigma_{f,n}^2} \quad (15.5)$$

To enhance the capacity of the femtocell while safeguarding the adjacent MUEs, the CR technique is applied to the femtocell. The suggested method has two steps: sensing of the channel and scheduling of the resources. The spectrum sensing process is able to find which of the frequency chunks physical resource blocks (PRBs) are not occupied by the MUEs. If the PU is assigned with the PRB, then FUE immediately vacates the channel and waits till the completion of transmission of the PU.

The channel occupancy status can be known by using hypothesis testing. Let g_1 and g_0 be the hypothesis denoting the presence or absence of MUE transmission. The received signal of the FBS is given by

$$y(n) = g(n)x(n) + n_o(n) \quad (15.6)$$

where $g(n)$ is the channel gain from MUE to the FBS.

$x(n)$ is the macro user transmitted signal.

$n_o(n)$ is the AWGN noise with zero mean and unity variance.

The signal received by the nth sub-channel can be approximated using discrete Fourier transform (DFT),

$$Y_n = \frac{1}{\sqrt{N_{\text{DFT}}}} \sum_{i=1}^{N_{\text{DFT}}} y(n) e^{\frac{-2\pi i x}{N_{\text{DFT}}}} \quad (15.7)$$

$$Y_n = G_n X_n + N_0 \quad (15.8)$$

where n is the number of sub-channels, i.e. $n = 1, 2, 3N_f$ and G_n, X_n, and Y_n are the discrete frequency response of $g(n)$, $x(n)$, and $y(n)$, respectively. The transmitted signal $x(n)$, the channel gain G_n, and the additive white noise N_0 are independent of each other. By using hypothesis testing the occupancy of the channel can be determined given in equation (15.9).

$$Y_n = \begin{cases} N_0, \dots H_{0,n} \\ G_n X_n + N_0, \dots H_{1,n} \end{cases} \quad (15.9)$$

where $H_{0,n}$ represents idle channel and $H_{1,n}$ represents a busy state of the channel 'n'. The signal energy (S_n) received to identify the status of the channel is given in equation (15.10).

$$S_n = \frac{1}{\tau}\sum_{t=1}^{\tau} \| Y_n(t) \|^2 \tag{15.10}$$

where τ represents time duration to detect the existence of signal in the channel or not. Assume η is the threshold level for signal detection.

a. If $S_n > \eta$, i.e. $H_{1,n}$, there is the presence of the signal, i.e. the channel is occupied.
b. If $S_n < \eta$, i.e. $H_{0,n}$, there is no signal, i.e. the channel is free.

The probability density function of the hypothesis is given by the gamma function, i.e. $\Gamma_n \cong \| G_n X_n \|^2$, where Γ_n does MUE measure the Signal to Noise Ratio (SNR) with respect to the nearby FBS.

The false alarm detection and probability is represented by

$$P_{f,n} = P(S_n < \eta \mid H_{o,n}) \tag{15.11}$$

The probability of detection is represented by

$$P_{d,n} = P(S_n > \eta \mid H_{1,n}) \tag{15.12}$$

The threshold level (η) in order to achieve the target SNR is defined by

$$\eta = \sqrt{\frac{2\Gamma_n + 1}{\tau}} Q^{-1}(P_{d,n}) + \Gamma_n + 1 \tag{15.13}$$

where $Q^{-1}(P_{d,n})$ is the inverse Q function probability of detection of signal.

The false alarm probability of detection is defined by equations (15.14) and (15.15):

$$Q_f = 1 - \prod_{i=1}^{N_f}(1 - P_{f,n}) \tag{15.14}$$

$$Q_d = 1 - \prod_{i=1}^{N_f}(1 - P_{d,n}) \tag{15.15}$$

where N_f is the number of frequency channels in each zone. $P_{f,n}$ and $P_{d,n}$ be the nth false alarm probability of frequency channel and probability of detection (N_f), respectively. The FBS identifies the sub-channel f_i which is having low energy level, i.e. ideal.

Figure 15.4 shows allocation of frequency channel to the cell edge UEs. At the boundary region of the area femtocells were deployed. The macro users are allocated with the sub-channels from MBS and femto users with the unoccupied channels.

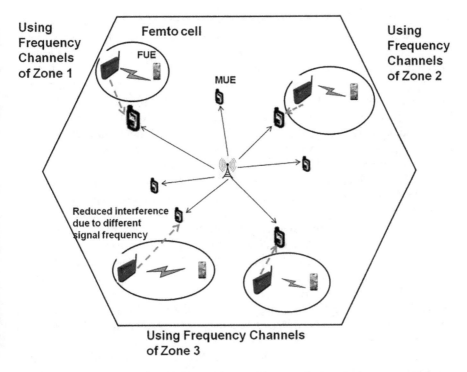

FIGURE 15.4 Allocation of frequency channel of the proposed method.

The interference level remains minimum as the frequency of both the channels was different. It designates N_f frequency channels for propagation. In j^{th} femtocell, the interference of the associated FUE for the nth sub-channel (where $n \in N_f$) is given by

$$\xi_j^n = \frac{\| g_i^n \|^2}{I_j^n}, n \in N_f \tag{15.16}$$

where $\| g_j^n \|^2$ is the channel's gain and I_j^n is the combined interference of associated FUE for frequency channel 'n' in the set of N_f channels. The FBS arranges N_f channels as per the measured $\{\xi_j^1, \xi_j^2, \ldots \xi_j^Z\} \in N_f$.

Then it assigns the minimum interference frequency channels between the FBS and FUE for communication purpose. The FBS has to dynamically regulate power of the assigned channels to minimize inter-tier interference between the femto cells. In this way the user receives high SINR, high throughput, better channel capacity, and guaranteed QoS. Moreover the spectrum efficiency enhances as per the specification of International Mobile Telecommunications (IMT)-Advanced (i.e. ≥ 1bps/Hz for the indoor scenario) [29].

15.6 RESULTS AND DISCUSSION

The performance of this research work is compared with the random resource allocation method. Figure 15.5 presents the cumulative distribution function (CDF) of

SINR of MUEs. The total number of MUEs was 200, and the number of femtocell clusters was 10. The FBS utilizes all the sub-channels present in the set of N_f number of operating channels. The transmitting power was dynamically varied to have minimum interference and maximum spectrum efficiency. The CFC allocates best quality channels to the FUEs to obtain guaranteed QoS with high SINR and better throughput. However at the same time the MUEs may not get better signal quality because of interference. The proposed method safeguards the MUEs with negligible deterioration of signal quality of FUEs. The femtocell clusters can manage this at the cell edge area. The SINR of cell edge MUEs reduces due to increased distance between the user and the MBS. At the cell boundary, the CFC clusters serve the cell edge UEs. The sub-channels used by the CFC were different than the macro cell used by the MUEs positioned at the cell center. The CFC senses the unoccupied frequency channels and assigns the FUEs to obtain better SINR and throughput. In this way, the interference reduces and the capacity of femtocell increases along with the spectrum efficiency of the network. This adds novelty to the proposed research work.

Figure 15.5 shows the CDF of SINR (in dB) of MUEs. It shows that the proposed resource allocation approach is 18% higher than the random resource allocation procedure. The number of MUEs investigated is 150 and the number of femtocell clusters is 25. The time interval of sensing a sub-channel was $\tau = 30$ m sec. The proposed cognitive sensing method yields nearly 98% accuracy of probability of detection with 9 dB SNR.

Figure 15.6 shows the CDF of SINR (in dB) of FUEs. The result indicates 21% higher performance in comparison with the random resource allocation approach. Figure 15.7 shows the enhancement of spectral efficiency (bps/Hz) in cognitive method (8 bps/Hz) compared to the random resource allocation (6 bps/Hz). The cognitive method is highly efficient for spectrum sensing, allocation, and utilization of the frequency channels in 5G HetNet. By using this method the interference can be minimized by adjusting the

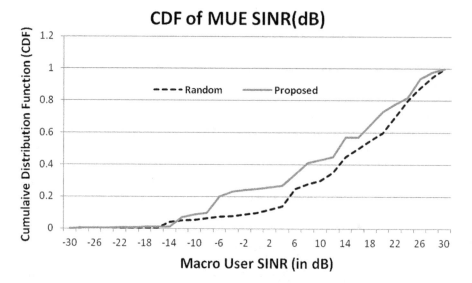

FIGURE 15.5   CDF of MUE SINR (in dB).

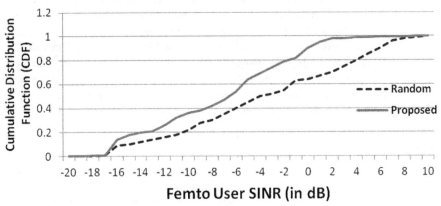

FIGURE 15.6 CDF of FUE SINR (in dB).

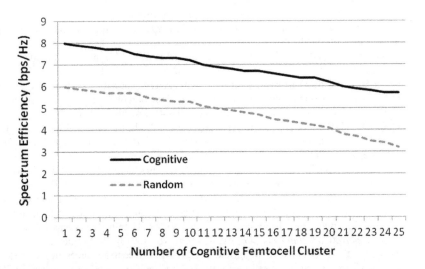

FIGURE 15.7 Spectrum efficiency of CFC cluster.

power level of the BSs. The radio resources will be allocated properly among the users and UEs will get guaranteed QoS. Furthermore, the space time-frequency RBs utilized adequately to increase the spectrum efficiency of the network. This adds significance to the research work in two-tier HetNet of 5G technology.

15.7 CONCLUSION

The main purpose of CFC is to escalate the network's capacity by serving the data-hungry indoor devices. The FFR technique differentiates the cell center from the cell edge area. The cell boundary areas served with cognitive empowered femtocells

(CFC), which were deployed in clusters. The frequencies used by the CFCs are different from the MBS (cell center), which mitigates inter- and intra-tier interference. It also provides an efficient solution for the spectrum scarcity problem by periodically scanning the frequency resources to know the occupancy status. The CR-enabled CFC opportunistically assigns licensed and unlicensed RBs to the user. The cell edge users are served with high signal strength by the CFC clusters, and hence the probability of outage reduces. It is concluded that the CFC minimizes interference with efficient spectrum allocation and guaranteed QoS. It also increases the spectrum efficiency of the 5G HetNet by proper allocation of the RBs.

REFERENCES

1. J. Mitola and G. Maguire, "Cognitive radio: Making software radio more personal," IEEE Personal Communications, 6:13–18, August 1999.
2. M. Carmen Lucas-Estañ, and J. Gozalvez "Mode selection for 5G heterogeneous and opportunistic networks", *IEEE Access*, vol. 7: 113511–113524, August 2019.
3. A. Gupta, A. Srivastava, R. Anand, and T. Tomažič. Business application analytics and the internet of things: The connecting link. In Gulshan Shrivastava, Sheng-Lung Peng, Himani Bansal, Kavita Sharma, Meenakshi Sharma (eds), *New Age Analytics: Transforming the Internet through Machine Learning, IoT, and Trust Modeling*, New York, 249–273, 2020.
4. K.-l.A. Yau, J. Qadir, C. Wu, M.A. Imran, and M.H. Ling "Cognition-inspired 5G cellular networks: A review and the road ahead", *IEEE Access*, vol. 6, 2018.
5. A. Al-Dulaimi, S. Al-Rubaye, and Q. Ni, "Energy efficiency using cloud management of LTE networks employing fronthaul and virtualized baseband processing pool", *IEEE Transactions on Cloud Computing*, vol 7(2): 403–414, April–June 2019.
6. D.-C. Oh, H.-C. Lee, and Y.-H. Lee "Cognitive radio based femtocell resource allocation", *International Conference on Information and Communication Technology Convergence (ICTC)*, November 2010, Jeju, South Korea.
7. D. Pliatsios, P. Sarigiannidis, S. Goudos, and G.K. Karagiannidis, "Realizing 5G vision through Cloud RAN: Technologies, challenges, and trends", *EURASIP Journal on Wireless Communications and Networking* vol. 2018: 1–15, Article number: 136, 2018.
8. M.F. Hossain, A. Uddin, M. Topojit, D. Farjana, B. Mosharrof, and K.Z. Islam, "Recent research in cloud radio access network (C-RAN) for 5G cellular systems - A survey" *Journal of Network and Computer Applications*, vol. 139: 31–48, August 2019.
9. M. Peng, Y. Sun, X. Li, Z. Mao, and C.g. Wang, "Recent advances in cloud radio access networks: System architectures, key techniques, and open issues", *IEEE Communications Surveys & Tutorial*, Vol. X(Y): 2282–2308, 2016.
10. I. Budhiraja, S. Tyagi, S. Tanwar, N. Kumar, and M. Guizani, "CR-NOMA based interference mitigation scheme for 5G femtocells users", *2018 IEEE Global Communications Conference (GLOBECOM)*, Abu Dhabi, UAE, Dec. 2018.
11. Cisco Annual Internet Report (2018–2023) White Paper. https://www.cisco.com/c/en/us/solutions/collateral/executive-perspectives/annual-internet-report/white-paper-c11-741490.html
12. X. Wang, P.-H. Ho, A. Wong and L. Peng, "Cognitive-empowered femto cells: An intelligent paradigm for femtocell networks" *Wireless Communications and Mobile Computing*, vol. 2018, Article ID 3132424, 9 pages, Doi: 10.1155/2018/3132424.
13. L. Huang, G. Zhu, and X. Du "Cognitive femtocell networks: An opportunistic spectrum access for future indoor wireless coverage" *IEEE Wireless Communications*, vol. 20(2): 44–51, 2013.

14. D. Gale and L. S. Shapley, "College admissions and the stability of marriage", *The American Mathematical Monthly*, vol. 69(1): 9–15, 1962.
15. L. Zhang, T. Jiang, and K. Luo, "Dynamic spectrum allocation for the downlink of OFDMA-based hybrid-access cognitive femtocell networks", *IEEE Transactions on Vehicular Technology*, vol. 65(3): 1772–1781, March 2016.
16. S. Al-Rubaye, A. Al-Dulaimi, and J. Cosmas, "Cognitive femtocell" *IEEE Vehicular Technology Magazine*, vol.6(1): 44–51, 2011.
17. G. Gur, S. Bayhan, and F. Alagoz, "Cognitive femtocell networks: An overlay architecture for localized dynamic spectrum access" *IEEE Wireless Communications* 17(4): 62–70, 2010.
18. G. Zhang, X. Ao, P. Yang, and M. Li "Power management in adjacent cognitive femtocells with distance-dependent interference in full coverage area" *EURASIP Journal on Wireless Communications and Networking* vol. 2016: 1–10, Article number: 8, 2016.
19. O. Grøndalen, and M. Lähteenoja, "Business case evaluations for LTE network offloading with cognitive femtocells", *Telecommunications Policy*, vol. 37(2–3): 140–153, March–April 2013.
20. J. Ghosh, and S.D. Roy, "The implications of cognitive femtocell based spectrum allocation over Macrocell networks", *Wireless Personal Communications: An International Journal*, vol. 92(3): 1125–1143, February 2017.
21. A. Salman, I.M. Qureshi, S. Saleem, and S. Saeed, "Spectrum sensing in cognitive femtocell network based on near-field source localization using genetic algorithm", *IET Communications*, vol. 11(11): 1699–1705, 03 August 2017.
22. H.M. Elmaghraby, and Z. Ding, "Scheduling and power allocation for hybrid access cognitive femtocells" *IEEE Transactions on Wireless Communications*, vol. 16(4): 2520–2533, April 2017.
23. A.F. Snawati, R. Hidayat, S. Sulistyo, and W. Mustika "A comparative study on centralized and distributed power control in cognitive femtocell network", *2016 8th International Conference on Information Technology and Electrical Engineering (ICITEE)*, IEEE, Yogyakarta.
24. S. Godase, and T. Wagh, "Dynamic spectrum allocation for hybrid access cognitive femtocell network", *International Conference on Information, Communication, Instrumentation and Control (ICICIC)*, 17–19 August 2017.
25. I. Al-Samman, R. Almesaeed, A. Doufexi and M. Beach, "Heterogeneous cloud radio access networks: Enhanced time allocation for interference mitigation", *Wireless Communication and Mobile Computing, Hindawi Publications*, vol. 2018: 1–16, Article ID 2084571, 2018.
26. J. Wu, Z. Zhang, Y. Hong, and Y. Wen, "Cloud Radio Access Network (C-RAN): A primer" *IEEE Network*, vol. 29(1): 35–41, January/February 2015. doi: 10.1109/MNET.2015.7018201
27. T. Salman, "Cloud RAN: Basics, advances and challenges, a survey of C-RAN Basics *Virtualization, Resource Allocation, and Challenges*", 1–16, April 17, 2016.
28. G. Sahu, and S.S. Pawar, "An approach to reduce interference using FFR in heterogeneous network", *SN Computer Science, Adaptive Computational Intelligence Paradigms and Applications* vol. 1: 1–6, Article number: 100 (2020), July 2019.
29. G. Sahu and S. Pawar. "An approach to enhance energy efficiency using small cell in smart city". *Proceedings 2019: Conference on Technologies for Future Cities (CTFC)*, December 24, 2018. SSRN: https://ssrn.com/abstract=3349626, http://dx.doi.org/10.2139/ssrn.3349626.

Index

ABI 244
absolute blank subframe (ABS) 254
access 220
according 217
accuracy 78
ace 218
acoustic 4, 7, 10
adaptive 226
AES 17–18, 20–22
alert messages 79
alienated 219
alliance 217
allocated 219
allocation 217, 218
analysis 217, 252
analyst 226
analytics 21, 82, 20, 22, 22, 24, 226
APCC 211, 212, 213, 214
applications 50, 223, 226
apprehensions 218
approach 226
architecture 218
assigned 218, 223
assigning 218
assisting nodes/auxiliary nodes 174, 183, 184
asymmetric cryptography 21
atmosphere 218, 223
authentication 17–19, 204, 209, 210, 211, 213
automatic 226
automation 217
availability 219
available 218, 219, 220
average 224, 226
AWGN noise 261
AWS account 131
AWS console 131

background services 254
balancing 218, 219, 221, 223, 226
bandwidth 106, 110, 120, 121
base 223
base band unit (BBU) 253
based 217, 220
basically 218, 223
beans 224
benefactors 218
big data 67
biometric 126
biometric verification 19–20
Bitcoin 191, 195, 196, 197

block chain 17, 19–20, 24, 25, 27, 126, 191, 192, 199, 200, 230–232, 235, 237, 239, 243, 245, 247, 248
bodily 220, 221
broadest 218
broker 219
broker services 36
burst 226
business 217
busy 219

CaaS 68
calculate 220, 222
capability 217
capacity 220, 221, 264
case 219, 220, 222, 226
challenges 218, 226
channel 252
classification 70
classifier 128
clients 218
cloud 217, 218, 220, 222, 223
cloud computing 16, 34, 66, 203, 205, 207, 214
cloud environment 16, 19–20, 24
cloudlets 223, 226
cloud overloading 206, 209, 211, 215
cloud providers 127
cloud RAN (C-RAN) 253, 254
Cloud Service Provider (CSP) 16, 18, 20, 24–25
cloud storage 16–17, 19, 21
CloudSim 217, 223, 224
cluster 259
Cluster-based Underwater Wireless Sensor Network (CUWSN) 9
cluster head 174, 177, 183, 184, 185
clustering 70
cognitive cycle (CC) 252
cognitive FAP 259
cognitive femtocell (CFC) 251, 254
cognitive radio 251, 252
college 217
commercial 226
communication 226
community 217
compare 219
complete 220, 222
comprising 223
computation 219, 220
computational 217, 219
computing 217, 218, 219, 220, 223

269

concentrating 223
conclusion 217
concurrently 220
condition 226
conference 226
confidentiality 16–17, 20
content prefetching 74
co-operative sensing 257
coordinated multi point (CoMp) 254
core 220, 221
corresponding 219, 220
cost efficiency 252
co-variance detection 257
COVID 217
cryptographic algorithm 17
cryptography 16, 21
CSP *see* Cloud Service Provider (CSP)
CST 17, 20–22
CUDA 128
cumulative distribution function (CDF) 263
currently 219
cybercriminal 194
cyclostationary detection 257

data 218, 220, 222, 224, 226
data aggregation 74
data integrity 16–17, 19, 24–25
data matrix code 17, 19, 25
data owner 18–20, 24–25, 27
data transaction 18–19, 27
DCB 219, 221, 222, 223
dealing 223
deciding 218
decreased 226
decryption 16, 21–22, 26
delay 88, 92–95, 98–100
depends 219, 220
deploying 223
DES 17–18, 20–22
descending 220
design 218
designated 219
determine 220
developers 217
different 218, 223
difficult 220
digital twins 52
digitizer 128
discrete fourier transform (DFT) 261
distributed 217, 218, 226
distribution 223
dynamically 218

EC2 135
effectiveness 223
e-healthcare 189, 193, 196, 197, 199
elasticity 223

electricity transmission 49, 52
Emrod 49, 63, 64
encryption 16, 21–22, 26
energy consumption 174, 183, 184, 185
energy cost 53, 54
energy data centers 51–54
energy detection 257
energy efficiency 252
ETH 234
Extended File Hierarchy Attribute-Based Encryption 18

facility 217, 218
factors 219
false alarm detection 261
FANETs 142, 162
feature vector 127
femto base station (FBS) 261
femtocell 252
femtocell access point (FAP) 255
File Remotely Keyed Encryption and Data Protection (FREDP) 18
finish 217, 219, 221, 223, 225
flexibility 51, 217
flying ad-hoc network 32
fog computing 32, 34
foremost 218
forthcoming 226
fractional frequency reuse (FFR) 254
FREDP *see* File Remotely Keyed Encryption and Data Protection (FREDP)
frequency 3, 10, 11, 12
fronthaul 254

Gale-Shapley spectrum sharing 254
gamma function 261
geography 50
graphical processing unit 128

hadoop 129
handling 220
HANET 104, 105, 107, 119, 120
hash function 24
hash key 126
hash tag 24–25
HBase 129
healthcare 230, 232, 235, 236, 237, 238, 239, 244, 248
Heroku 17
Heterogeneous Cloud Radio Access Network (HCRAN) 254
heterogeneous network (HetNet) 251
honor 252
host 220, 221
HTML 223
human identification 126
hybrid cloud 205

Index

hybrid cryptographic technique 17, 19, 21
hypothesis 223
hypothesis testing 261

ID 219, 223
IDE 223
imitation 223, 226
imperative structure model 238
initially 221
innovation leadership 60
inscribed 223
instance 222, 223
instructions 220, 226
integrated 223
integrity *see* data integrity
interactive services 254
internet 217, 218
Internet of Biometric Things (IoBT) 126
Internet of Things 190, 198
interweave 255
investigation 223
iris verification 17, 20, 27

JSON 244

K Nearest Neighbourhood 128

leadership in innovation 60
learning 252
Linux 223
load 217, 218, 219, 221, 223

MAC 105, 121
macro base station (MBS) 252
mainframe computer 16
MANET 85–86, 88, 98–101
Markov, Gauss 85, 100
mesh networks (MSN) 51
methodology 217, 224
Microsoft 223
millimeter wave (mmWave) 253
million 226
minimum 222, 223
MIPS 226
mobile ad-hoc networks 85, 100–101
mobile nodes 174, 183
mobility 104, 105, 199, 120, 121, 142, 143, 144, 145, 146, 147
mobility models 85, 86, 88, 99–101
model 218, 226
modeling 226
motto 217
mounting 223
multidimensional technique 127
multi-hop 142, 143, 147, 151, 155
MWNs 142, 143, 148, 149, 151

Netbeans 223
network 51
network function virtualization (NFV) 252
networking 218
node 218, 231, 232, 233
NOSQL database 128
novel 223

online 218
open 226
operating 223
operatives 217
opportunities 226
optical fiber 253
orthogonal frequency division multiple access (OFDMA) 254
outcome 217, 223
overlay 255

package 218
pandemic 217
parallel 217
parameter 220
pathloss 259
PDR 88, 95–100
period 218, 219
peripheral 218
physical resource blocks (PRB) 261
pilot-based coherent detection 255
platform 218, 223
podiums 223
poisson point process (PPP) 255
privacy 205
private broadband networks 52
probability of detection 261
problem 219
procedure 218, 221, 222, 225
process 218, 219
processing 217, 221, 224, 226
program 219
programmers 223
programming 223
progress 220
project 226
projected 220, 221
prototype 218

QoS 85, 88, 98–100, 252

radio frequency identification 195
random waypoint 86, 101
range 50
Re'seau de Transport d'E'lectricite' (RTE) 49
realtime call 254
registration phase 18, 20
reliability 218

remote radio head (RRH) 254
resource allocation 254
resource blocks (RBs) 252
resource slicing 252
RTE 49
running 224

6v'S 67
scaling 226
scheduling 21, 92, 21, 226
SDARP 173
SDG-7 49
security 16, 19, 23, 204, 205, 206, 207
segmentation of WPT 50
selected 219, 223
selection 219
sensing 252
sensor networks 51
sensors 173, 177–183, 185
servers 218
service 217, 218, 219, 222
service overloading 204, 206, 208, 210
service-oriented network (SON) 208
setup phase 18, 20, 24
simulation 144, 145, 146, 148, 163, 218, 226
smart-health 237
social 218
society 226
SOFM 214, 215, 216
software 217, 218, 223
software defined radio (SDR) 251
sort 220
source 226
spectrum efficiency 258
spectrum sharing 251
spell 219
spending 219
SS 219, 220, 221
stages 221
standalone sensing 257
static nodes 174, 183, 184
status 219, 221, 223
storage 217
structure 220, 223
subcarrier 259
sustainable development goals (SDG) 49
symmetric cryptography 21
system 218, 226

task 218, 219, 221
TCP/IP 105, 121

technology 217, 218
technology segmentation of WPT 50
terms 220
test 218, 223
testing 218
third party 16–19, 25
Third Party Administrator (TPA) 19, 24–25, 27
throughput 88–92, 98–100
timestamp 24
toolkit 226
total processing time (TPT) 204
TPA *see* Third Party Administrator (TPA)
transmission 51
trust 199
TS 219, 220, 221
tuning parameter 86, 88–93, 95–100
turnaround 217, 224, 225

UAV based fog 35, 36
ultra high frequency (UHF) 258
UN Sustainable Development Goals 49
underlay 255
underutilization 219
underwater 2, 3, 4, 5
underwater Internet of Things 8
underwater wireless optical communication 4
unmanned aerial vehicles 32

VANET 32, 34, 104, 105, 107, 117, 118
very high frequency (VHF) 258
virtual 218, 219, 220, 221
virtualization 218
virtual power transmission 49, 50, 52–54, 59
virtual twins 52
VM 218, 219, 220, 221, 222
VMk 221
VMs 218, 219, 220, 221, 222
VPT 49, 54, 59, 60

wi-fi long and short 61
wire free worry free 51
wire-free 51
wireless (cordless)-local-area-network 204
wireless mesh network (WMN) 51
Wireless Power Transmission (WPT) 49–56, 58–63
WLAN 205
WMN 104, 105, 107, 118
World Health Organization 190, 195
worry-free 51
WPT *see* Wireless Power Transmission (WPT)
WSN 142, 146, 147, 148, 149, 152, 155, 162

CPSIA information can be obtained
at www.ICGtesting.com
Printed in the USA
BVHW052215230522
637897BV00002B/35